AF454170

# A Review on Chironomids

# About the Author

Dr. Deepak Rawal (SET, CSIR-NET, Ph. D.) is an eminent zoologist working in the field of Entomology. He is working as an Assistant Professor in the Department of Zoology, Mohanlal Sukhadia University, Udaipur (India). He got his expertise on Chironomids during his Ph.D. work, till then he is working on Chironomids.

# A Review on Chironomids

Dr. Deepak Rawal

2015
**Scholars World**
*A Division of*
**Astral International (P) Ltd**
New Delhi 110 002

© 2015 AUTHOR

*Publisher's note:*

*Every possible effort has been made to ensure that the information contained in this book is accurate at the time of going to press, and the publisher and author cannot accept responsibility for any errors or omissions, however caused. No responsibility for loss or damage occasioned to any person acting, or refraining from action, as a result of the material in this publication can be accepted by the editor, the publisher or the author. The Publisher is not associated with any product or vendor mentioned in the book. The contents of this work are intended to further general scientific research, understanding and discussion only. Readers should consult with a specialist where appropriate.*

*Every effort has been made to trace the owners of copyright material used in this book, if any. The author and the publisher will be grateful for any omission brought to their notice for acknowledgement in the future editions of the book.*

*All Rights reserved under International Copyright Conventions. No part of this publication may be reproduced, stored in a retrieval system, or transmitted in any form or by any means, electronic, mechanical, photocopying, recording or otherwise without the prior written consent of the publisher and the copyright owner.*

**Cataloging in Publication Data--DK**
Courtesy: D.K. Agencies (P) Ltd. <docinfo@dkagencies.com>
**Rawal, Deepak, author.**
A review on chironomids / Dr. Deepak Rawal.
pages    cm
Includes bibliographical references (pages      ).
ISBN 9789351307426 (International Edition)
1. Chironomidae.   I. Title.
DDC 595.772    23

Published by         :  **Scholars World**
A Division of
**Astral International Pvt. Ltd.**
– ISO 9001:2008 Certified Company –
4760-61/23, Ansari Road, Darya Ganj
New Delhi-110 002
Ph. 011-43549197, 23278134
E-mail: info@astralint.com
Website: www.astralint.com

*Laser Typesetting*    :  **GRB 7Color Service,** New Delhi - 110 084

*Printed at*          :  **Thomson Press India Limited**

**PRINTED IN INDIA**

# Preface

When I was doing my Ph. D., my research area was related to Chironomids and I found that no book is giving comprehensive details about it and peoples are not taking  them seriously because of their non biting habits. Then I decided to write this book so in future there is a easy way to get information about Chironomids who wants to research in the field of Chironomids. This book is based on published papers related to research area of Chironomids. Each paper is thoroughly studied and interpreted by me, whose summary I am going to give in this book. Review in this book is arranged chronologically for convenience in readings to readers. Bibliography at the end of this book is according to APS style.

# Acknowledgement

I feel privileged to express my profound gratitude to my Research Guide, Dr. Preeti Singh, Assistant Professor, Department of Zoology, Mohanlal Sukhadia University, Udaipur, India., for his constant encouragement, inspiration, guidance and his dedication towards my work.

I also want to acknowledge my family specially my son Rishi. My appreciations are due to them for their precious and priceless contribution and support to me in reaching this milestone in my career.

*Dr. Deepak Rawal*

# Contents

*Chapter 1*
# INTRODUCTION

## 1.1 CHIRONOMIDS

Kingdom – Animalia

Phylum – Arthropoda

Class – Insecta

Order – Diptera

Suborder – Nematocera

Family – Chironomidae

The Chironomids, often called non biting midges or blind mosquitoes, are abundant and widespread aquatic insects. These insects spend the greatest part of their life cycle in larval form, occupying a wide range of habitats. Chironomids are holometabolous insects. Their life cycle includes three aquatic developmental stages (egg, larva and pupa) and a terrestrial reproductive stage (winged adult). Chironomid larvae have four intervals or instars between hatching from the egg and becoming a pupa, shedding its exoskeleton (molting) at the end of each instar. The largest, fourth instar is the most reliable stage for observing distinguishing features of the different genera and species. Ecdysone induces molting in Chironomids.

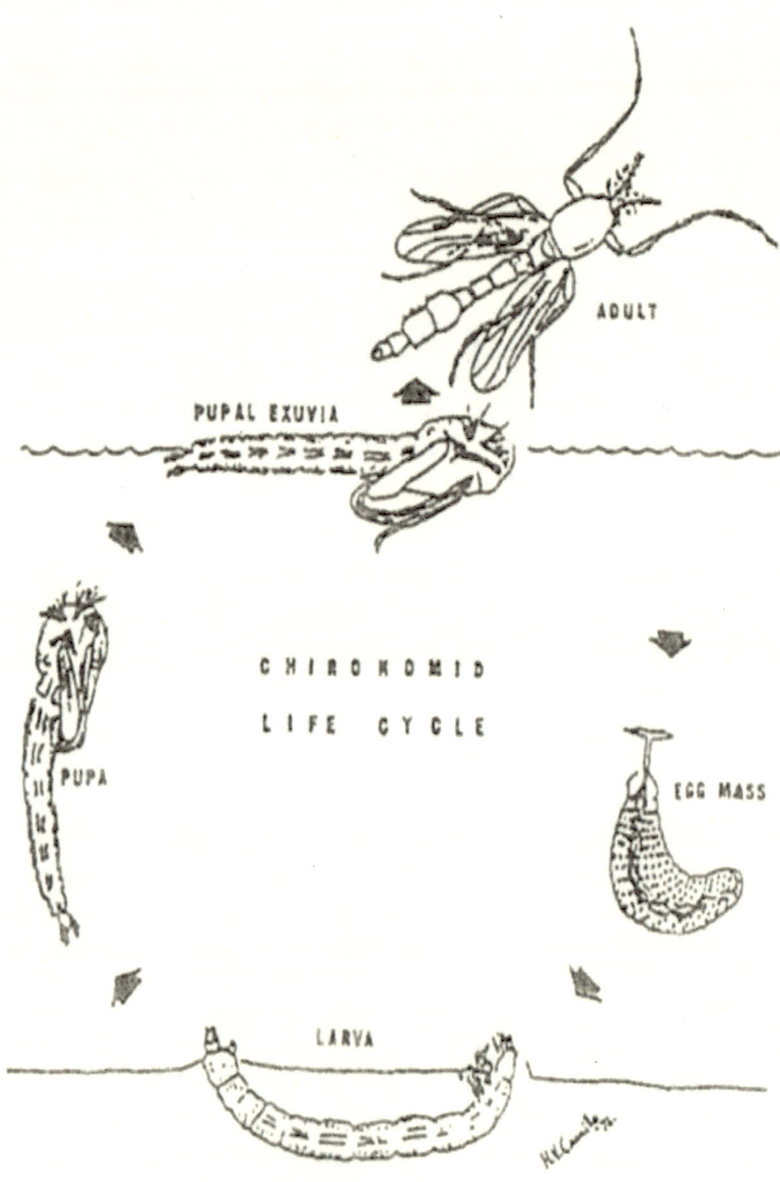


**Figure.1.1:** Chironomid life cycle.

Adult males are distinguished from females by the presence of plumose (feathery) antennae. The average life cycle of Chironomids is four weeks. Chironomids ranges in length from less than 1 mm to more than 2 centimeters. The duration and characteristics of each life stage are species specific. Chironomids spend most of their lives as benthic larvae living and feeding on the sediment (Oliver 1971). Chironomids are closely related to mosquitoes (Culicidae) and biting midges (Ceratopogonidae). Unlike Culicidae and Ceratopogonidae, female Chironomids do not bite due to absence of elongated mouthpart and they lack wing scales too. The larvae of most Chironomids hatch out from the eggs and usually lead bottom dwelling life forming tubes with the help of silk secreted from the salivary glands taking or clay particles and organic matters from the substratum. Some forms are found to live inside muddy substratum.

Chironomids are one of the most dominant, widespread and diverse aquatic invertebrate taxa in freshwater systems (Armitage 1995). Chironomids are distributed from the Arctic to the Antarctic and from the seas to permanent snowfields. Chironomids live in the glaciated areas of the highest mountains, including an elevation up to 5600 meters in the Himalaya (Kohshima 1984; Saether & Willassen 1987) and are active in a temperature of -16°C. Larvae of *Sergentia* live at over 1000 meters depth in the abyss of Lake Baikal (Linevich 1963). Chironomids are an important food source for large predatory invertebrates, fishes and birds (Winfield & Winfiels 1994; Hudson *et al.* 1995; Benke *et al.* 2001). Chironomids are important contributors of carbon and energy flow to higher tropic levels (Benke and Wallace 1997). Chironomids have an important role in aquatic food webs, representing a major link between producers, such as phytoplankton and benthic algae and secondary consumers (Tokeshi 1995).

It has been estimated that the number of species worldwide may be high as 15000. This high species diversity has been attributed to adaptation capability and ecological ability of larvae to extreme environmental conditions (Armitage *et al.* 1995). Some species can tolerate the most polluted conditions, whereas others have very special requirements and quickly disappear when their habitats are stressed. Therefore, the Chironomids possess exceptional potential to act as biological indicators of aquatic environmental health. The Chironomid larvae are found in all aquatic ecosystems, including freshwaters and seawaters. They are also semi aquatic and land dwellers. The larvae of most Chironomid species live on or in sediments, where they feed on organic matter (detritus) and associated macro fauna and flora. Because of their benthic habits, these larvae are directly exposed to contaminants in sediments throughout their development. Chironomids are often associated with degraded or low biodiversity ecosystems because some species have adapted to virtually anoxic conditions are dominant in polluted waters.

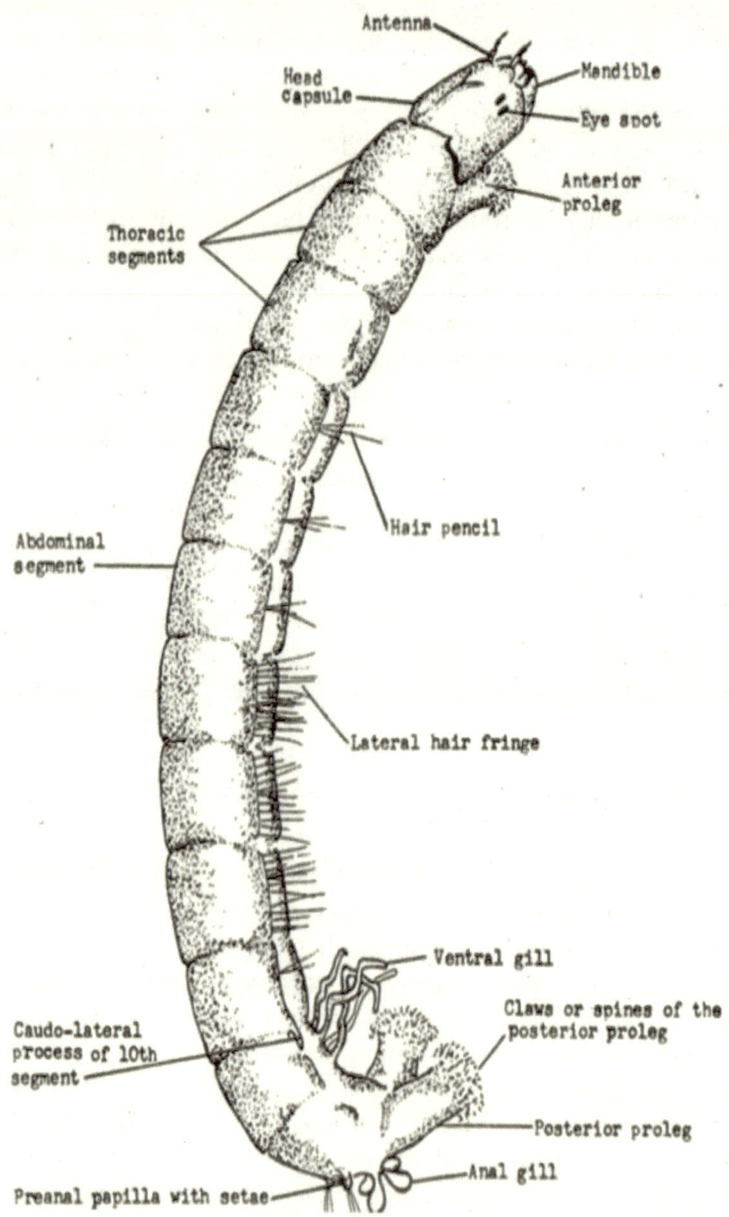


**Figure. 1.2:** Schematic diagram of Chironomid larva

Because of their close association with the benthic zone, easily identifiable life stages, ease of culture and sensitivity to chemical and environmental stressors, Chironomid larvae are commonly used as indicator species in laboratory and field-based toxicity tests (Lindegaard 1995; Environment

Canada 1997). They show both developmental (Timmermans *et al.* 1992; Dube & Culp 1996) and teratogenic (Hudson & Ciborowski 1996; Groendijk *et al.* 1998; Martinez *et al.* 2001) responses to a wide range of chemical contaminants. Chironomids are used for environmental monitoring. Distribution and occurrence of Chironomids indicate physicochemical characteristics of water bodies (Donald & Marry 1983). The deformities developed in the larval body parts as well as polymorphism developed in the polytene chromosomes of the larvae makes these small creatures useful tools for monitoring water quality in terms of ecological health.

There is a strong link between the Chironomid phenotype, specifically larval morphological deformities and chemical contaminants. Results from many field studies strongly indicate a relationship between increased incidence of deformations and toxic sediment stress (Hamilton & Saether 1971; Kohn & Frank 1980; Warwick 1980; Janssens de Bisthoven & Speybroeck 1994; Nazarova *et al.* 2001 etc.). Chironomid fossils are widely used by paleontologists as indicators of past environmental changes, including past climatic variability. Females of Chironomids deposit egg masses at the water's edge, each egg mass may contain 400-1000 eggs embedded in a thick, gelatinous matrix. The larvae of some species of Chironomids are red in color due to the presence of hemoglobin in the blood. Although Hemoglobin is widely distributed throughout the animal kingdom, its occurrence in invertebrates is restricted to only a few representatives of the large taxonomic groups.

The Chironomid Hemoglobin, members of insect respiratory proteins have been extensively studied and found to be very attractive research materials as they offer the simplest model for more complex Hemoglobin. In the past the ability of many Chironomid larvae to survive in conditions of low oxygen concentration has attracted attention and this ability has generally been considered to be related to their possession of hemoglobin. Their ability to capture oxygen is further increased by their undulating movements. Hemoglobin is present in all Chironominae, certain larval Tanypodinae and *Propsilocerus* and *Tokunagayusurika* amongst the Orthocladiinae.

Larvae of Chironomidae can cope with low oxygen concentration, increased salinity, wide range of pH and are above all, remarkably tolerant to organic and heavy metal pollution. Some scientists suggested that this is due to genetic adaptation. Chironomids are so small that they cannot be studied with unaided eyes, the species diversity is very high and the literature one has to consult for detailed identifications often needs numerous specialized works rather than just a few comprehensive ones. These and other reasons have combined to give this group an image of requiring relatively high effort of determinations and interpretations. For these reasons many macro-invertebrate workers have stayed away from adequate consideration of non biting midges. In most studies the importance of the group has been ignored due to taxonomic problems related to their identification. Commercial rearing of Chironomids has been attempted in several countries *viz.*: Hong Kong, Thailand as food for fishes.

## 1.2 POLYTENE CHROMOSOMES IN CHIRONOMID LARVAE

In the past, the only thing which made Chironomids famous was polytene chromosomes, which were originally observed in the larval salivary glands of *Chironomus* midges by Balbiani in 1881. Polytene chromosomes are present in several dipterans tissues, including the salivary glands, malpighian tubules and the epithelium of midgut and hindgut (Staiber & Behnke 1985; Michailova 1989). Polytene chromosomes also called giant chromosomes are very large and visible using a compound microscope. When properly stained, the alternating light/dark banding pattern of heterochromatin is visible. The banding pattern is species specific and permits the identification of reproductively isolated species (Martin 1979; Michailova 1989).

Chironomid larvae typical of Diptera possessing giant polytene chromosomes that forms when chromatin strands replicate but fails to divide (Case & Daneholt 1977). When genes on these chromosomes undergoes transcription the normally condensed chromatin strands uncoil and puffs appears. Puffs are sites of RNA synthesis and the rate of transcription is proportional to the size of the puff (Pelling 1964; Deneholt *et al.* 1969). Most Chironomid larvae have four banded (2n=8) polytene chromosomes whose standard banding pattern has been mapped by Hagele (1971) and Kiknadze *et al.* (1991). Polytene chromosomes are suitable structure for direct visualization of gene action (Beermann 1952; Breuer and Pavan 1953). Polytene chromosomes provide an excellent system for studying the structural organization of chromosomes because of their diameters. The bands of polytene chromosomes are considered as genetic markers to analyze divergence patterns of the linear genome structure during evolution.

The nucleolar organizer region (NOR), visible as an especially large puff, is an indicator of sublethal stress (Bentivegna & Cooper 1993; Hudson & Ciborowski 1996). It shows a marked decrease in size (Aziz *et al.* 1991; Michailova *et al.* 1998) and transcriptional activity (Planello *et al.* 2007), when larvae are exposed to contaminated sediments or elevated levels of trace metals. Larvae of fourth instars have a good type of polytene chromosomes observed in the salivary gland, used as tools for taxonomy and phylogeny of Chironomidae (Martinez 1979). Study of Chironomid polytene chromosomes is a part of the syllabus in all universities of India.

*Chapter 2*

# A REVIEW ON CHIRONOMIDS

Polytene chromosomes were originally observed in the larval salivary glands of *Chironomus* midges by Balbiani in 1881. Since then Chironomids are used as model organisms for the study of chromosomes. In the past the ability of many Chironomid larvae to survive in condition of low oxygen concentration has attracted attention and this ability has generally been considered to be related to their possession of hemoglobin. This view was however, been based on insufficient experimental evidences. Zavrel (1920) is of the opinion that the hemoglobin acts as an oxygen store to be utilized by the animal in periods of oxygen shortage. Leith (1916) had experimentally showed that this view is incorrect for the hemoglobin can only carry enough oxygen to the last the animal about 12 min. at 20°C.

The function of hemoglobin in *Chironomus thummi* larvae had been studied experimentally by Harnisch (1936), making use of the carbon monoxide method. The respiration of untreated animals was compared with that of animals in which oxygen carriage by the hemoglobin has been prevented by treatment with carbon monoxide. Any differences in respiratory rates found should therefore give a measure of the amount of oxygen carried by the pigment. Harnisch had studied both normal animals (O2 animals) and animals which have previously been subjected to severe oxygen lack (N2 animals). He found that in the former case at the air saturation the hemoglobin is not used in oxygen transport, but comes into action only at low oxygen concentrations. In the case of N2 animals, the rate of oxygen consumption is higher than that of normal animals and he concluded that at 17°C, the hemoglobin of *Chironomus* larvae from well aerated water does not function in oxygen transport at air saturation, but only at oxygen transport at oxygen pressure below 3 cc per liter. In his experiments he used micro-winkler method described by Fox & Wingfield (1938) for oxygen content determination.

The study on base composition of nucleic acids in chromosomes, puffs, nucleoli and cytoplasm of *Chironomus* salivary gland cells was done by Edstrom & Beermann (1962). The base composition of RNA from individually isolated giant chromosomes, puffed chromosome segments, nucleoli and samples of cytoplasm from *Chironomus* salivary gland cells was determined by microelectrophoresis. Data on the adenine: guanine quotient of the chromosomal DNA was also obtained. The results show that: 1) Chromosomal, nucleolar and cytoplasmic RNAs differ significantly from each other in base composition. 2) Nucleolar and cytoplasmic RNAs in spite of the difference, show great similarities with regard to the base composition and are both rich in adenine and uracil. 3) The RNA extracted from chromosome I differs significantly from the RNAs extracted from different segments of chromosome 4 and the latter differ significantly from each other. 4) The value for the RNA:DNA quotients of chromosome active genes, so called Balbiani rings. 5) The chromosomal RNA does not show a base symmetry in any of the investigated cases, nor is the content of guanine+cytosine, the same as that for DNA.

The demonstration of genic sex determination in *Chironomus* was first made by Beermann (1955) in his report of male limited inversion sequences in several populations of *Chironomus tentans* and *Chironomus pallidivitattus*. Different populations of *Chironomus tentans*, possibly representing geographically isolated races, have two differentiated genic mechanisms of sex determination involving either a dominant male determining factor in the left arm of chromosome 1 or a dominant female determining factor at the right tip of chromosome 1. In crosses between these populations, the male determining factor is epistatic to the female determining factor. No evidence of intersexuality has been found in such crosses. For determination of sex with chromosomal constitution, last instar larvae were examined by their developing gonads as described by Wuelker & Goettz (1968). It is especially interesting that in the combination of strong male and female factors, sexuality is male and fertility is essentially normal.

Unlike vertebrates in which the phenotypic expression of hemoglobin is limited, individual members of the genus *Chironomus* display a striking hemoglobin polymorphism. This hemoglobin multiplicity is both stage and species specific and can be effectively displayed by gel electrophoresis (Manwell 1966; Thompson & English 1966; Trichy 1966; Braunitzer *et al.* 1968; Wulker *et al.* 1969; English 1969; Laufer and Poluhowich 1971; Schin *et al.* 1974). Hemoglobin was first seen in the second instar larvae of *Chironomus* and may comprise 90% of the hemolymph proteins in the fourth instar (English 1969).

Bergtrom *et al.* (1976) found the fat body, as major site of hemoglobin synthesis in *Chironomus thummi*. In their experiments, fourth instar larvae of *Chironomus thummi* were incorporated with labeled amino acids *in vivo* and in organ culture. The products secreted into the hemolymph or into the culture medium were examined by acryl amide gel electrophoresis. Nine eletrophoretic bands were resolved as hemoglobins without staining. The results of culturing isolated salivary glands, gut and fat body demonstrated that the fat body is the major site of hemoglobin synthesis and secretion. Fat body which is mesodermal in origin synthesizes both heme and the globin of *Chironomus* hemoglobins. The gut tissue including the malpighian tubules also synthesize and secrets a small quantity of hemoglobin (about 5% of the total secreted by both gut and fat body from the same animal *in vivo*).

Manwell (1966); Wulker *et al.* (1966); Laufer & Polyhowich (1971) and others have demonstrated that the electrophoretic patterns of larval hemoglobin change with development. These changes are temporary related to molting and metamorphosis and may therefore be under the regulatory control of molting hormones (*e.g.* - Ecdysone). Although the fat body of higher Diptera (*e.g.* - Drosophila) undergoes histolysis at metamorphosis and is replaced by a new organ in the adult, this may not be the case in lower dipterans such as *Chironomus* and *Aedes* (Wigglesworth 1949; Trager 1937). Larval and adult *Chironomus* fat body is one and the same organ with a homogenous cell population, which functions to secrete two major protein products, hemoglobins and vitellogenins, each at a different stage in the life cycle of the insect.

Arndt-Tovin *et al.* (1983) confirmed the presence of left handed Z-DNA in bands of acid fixed polytene chromosomes in *Chironomus thummi* and *Drosophila melanogaster*. In their experiments, antibodies to DNA in the left handed confirmation bind to acid fixed polytene chromosomes of both *Chironomus thummi* and *Drosophila melanogaster*, as shown by direct and indirect immunofluorescence. Comparison of the phase contrast, immunofluorescence and DNA staining patterns show a predominant localization of the antibody to the regions of high contrast and DNA density. An alternate statement is that the left handed Z-DNA confirmation is present at a frequency of 0.02-0.1%. The measured differences reflected variations in the local density of Z-DNA sites and not in the affinity for the specific antibody, which appears to be relatively constant throughout the chromosomes. These observations taken together with results of biophysical studies on the properties of Z-DNA in solution suggested that regions of DNA in the left handed confirmation could be involved in higher order structural organization of chromosomes and possibly in modulation of their functional state.

Ashe *et al.* (1987) did the work on review of information on zoogeographical distribution of Chironomidae at subfamilies and genera level. The seven zoogeographical region *viz.*: Antarctica, Australoasian, Afrotropical, Nearctic, Neotropical, Oriental and Palearctic includes ten subfamilies of Chironomidae *viz.* Aphroteniinae, Buchonomyiinae, Cheilenomyiinae, Chironominae, Diamesinae, Orthocladiinae, Podonominae, Prodiamesinae, Tanypodinae and Telmatogetoninae. The family Chironomidae is a cosmopolitan group of dipteran insects representatives of which occur in all zoogeographic regions of the world. There are few areas where the Chironomidae can be regarded as being absent. The immature stages of most species occur in freshwater but many terrestrial, marine and brackish water species are known. Ashe (1983) recognized 355 valid genera within the ten Chironomid subfamilies. Some of those genera, though technically valid are of uncertain status. A few of the valid genera have since been synonymized with other genera and some new genera have been recently described. Representatives of six subfamilies Telmatogetoninae, Tanypodinae, Podonominae, Diamesinae, Orthocladiinae and Chironominae have been recorded from all zoogeographical regions. The remaining four subfamilies appear to have a more restricted distribution. The Chilenomyiinae are known from the Neotropical region, the Buchonomyiinae from Palearctic and Oriental, the Aphroteniinae from the Neotropical, Afrotropical and Australoasian regions and the Prodiamesinae from the Palearctic, Nearctic, Neotropical and Oriental regions. The Chironomid fauna of the Oriental is the least well known of all the zoogeographical regions. The adults of many described species are poorly and incompletely illustrated and described so that recognition of many described or even new species is very difficult. The immature stages have been largely ignored and are mostly unknown or undescribed. A catalogue of the Chironomidae of the Oriental region has been published (Sublette & Sublette 1973). It is evident from this catalogue that many of the species described from this region by Kieffer, now regarded as *nomina dubia* require re-examinationand and re-description in line with currently accepted diagnostic and illustrative methods. In recent years Chaudhuri and co workers have begun working on taxonomy of the fauna of the Indian subcontinent. Only two of the ten subfamilies (Chilenomyiinae & Aphroteniinae) have not been recorded from this region and there are at present only four genera *viz.*: *Neopodonomus, Asclerina, Eusmittia* and *Trichotendipes* which are unique to this region. Oriental region possess a fauna of 348 valid species (compiled from the catalogue of Sublette & Sublette 1973, Zoogeographic Record and various individual taxonomic records). Chironominae, Orthocladiinae and Tanypodinae are the dominant groups with

181 (52%), 78 (22.4%) and 67 (19.2%) species respectively from Oriental region. Representatives of some genera (*e.g.* - *Tanypus*) have been reported from hot springs at temperature up to 44.5°C (Thienemann 1954). *Parachironomus* species are known to be parasitic in Gastopods with *Parachironomus varus* Goetghebuer ectoparasitic in *Physa frontinalis* Linnaeus and *Parachironomus varus* var. *limnaei* Guibe reported as endoparasitic in *Limnaea limosa* (Guibe 1942). *Demeijerea* species mine in and feed on freshwater sponges and Bryozoa (Neff & Benfield 1970) and some species of *Xenochironomus* mine sponges (Pinder & Reiss 1983). *Polypedilum fallax* (Joh.) is reported to be parasitic and predatory on Trichopteran pupae of the genus *Potamophylax* (Otto & Svensson 1981) and *Collartomyia* also appears to behave similarly (Borkent 1984). *Dicrotendipes peringueyanus* Kieffer in tropical Africa is phoretic on freshwater crabs (*Potamonautes*) and was never found free living (Disney 1975).

Rovira *et al.* (1993) found a repetitive DNA sequence associated with the centromeres of *Chironomus pallidivittatus*. A clone containing centromere-associated DNA from *Chironomus pallidivittatus* was obtained by microdissection and microcloning. It hybridizes to the centromeric end of one chromosome and exclusively to regions in the three remaining metacentric chromosomes to which centromeres have previously been localized on cytological grounds. In the metacentric positions the hybridization can be assigned to thin bands. The clone contains 155 bp tandem repeats and short flanking regions presented in all of the centromeres. Titration experiments show that the four centromeres together contain 200 kb of 155 bp repeat per genome. Each repeat contains two invert repeats surrounding a region containing only AT base pairs, a feature with some similarity to functionally essential elements in the Saccharomyces cerevisiae centromere.

Postam *et al.* (1994) studied the interacting effects of cadmium toxicity and found limitation on the midge *Chironomus riparius* during chronic exposure in laboratory experiments. If the food was supplied *ad libitum*, both larval developmental time and mortality of the larvae were negatively affected by cadmium concentration of 2.0-16.2 µg/L. The number of eggs deposited per female and the mean life span of the imagines were not affected by cadmium. Integration of these separate effects into a population growth rate showed a cleared reduction with increasing cadmium concentration.

Belikov and Wieslander (1994) found that chromatin structures in salivary gland cells of *Chironomus tentans* have improved preservation under 70% ethanol fixation.

Caras (1996) studied the effect of diet quality on growth and development of recently hatched larvae of *Chironomus* gr. *plumosus*. He found that benthic algae derived detritus proved to be the food of higher quality producing a higher rate of growth than did the CPOM (coarse particulate organic matter) diet. In the

FPOM (fine particulate organic matter) diet no significant growth was obtained in the experiment. CPOM diet was leaf debris in the form of leaf disks of 1.5 cm diameter and FPOM diet was in the form of leaf particles of size less than 250 µm. Casas used STATGRAPHICS (1994) software for all statistical analysis.

Dutta *et al.* (1996) collected the adults belong to five genera of *Hernischia* complex from Duars of the Himalayas of West Bengal (India). Four species *Cryptochironomus curryi* Mason, *Cryptochironomus ramus* Mason, *Dermicryptochironomus vulneratus* Zetterstedt and *Harnischia curtilamellata* Malloch are recorded for the first time from India and four species *Cryptochironomus acuminatus, Cryptochironomus gracilis, Harnischia minuta* and *paracladopelma diutinistyla* are described as new. A key to the Indian species of *Cryptochironomus* Kieffer is presented. The female of *Cryptochironomus subovatus* Freeman, hitherto unknown is described. One Japanese species *Demicryptochironomus chuzequartus* Sasa, 1984 is proposed as a synonym of *Demicryptochironomus vulneratus* (Zetterstedt, 1838). The materials were preserved in ethanol (90%). Specimens were mounted following the technique of Das and Gupta for examination of external morphology. Types and specimens deposited in the National Zoological Collection, Calcutta.

Morcillo & Diez (1996) studied the effect of heat shock in *Chironomus thummi*. They found that on elevated temperatures, *Chironomus thummi* polytene chromosomes induce a giant Balbiani ring like structure at the right telomere of chromosome III, which is called T-BR III. Besides the T-BRIII other telomeres may appear additionally puffed but in a rather invariable way. The temperature necessary to induce T-BRs ranges from 33°C to 37°C, while the larvae are cultivated at 18°C in the laboratory. T-BR formation involves transcriptional activity, which could be detected by different methods. The local labeling all over the induced T-BR after a short pulse with [³H]-uridine undoubtedly means transcription at this structure under heat shock. T-BRs show similarities with the 93D locus in *Drosophila*.

The phylogeny of the subfamilies of Chironomidae is cladistically analyzed using parsimony by Saether (2000). Parsimony analysis was performed using PAUP 3.1.1 (Swafford, 1993) and MacClade 3.06 (Maddison and Maddison, 1996) on a Power Macintosh 8200/120. Cladograms were compared using MacClade 3.06. Preferred cladogram is shown below:

Most likely all Chironomid subfamilies can be placed in one of three semifamilies Telmatogetoninae, Tanypodinae or Chironominae. The placements of both Chilenomyiinae and Usumbaromyiinae however are highly tentative and need confirmation by the morphology of the immature stage.

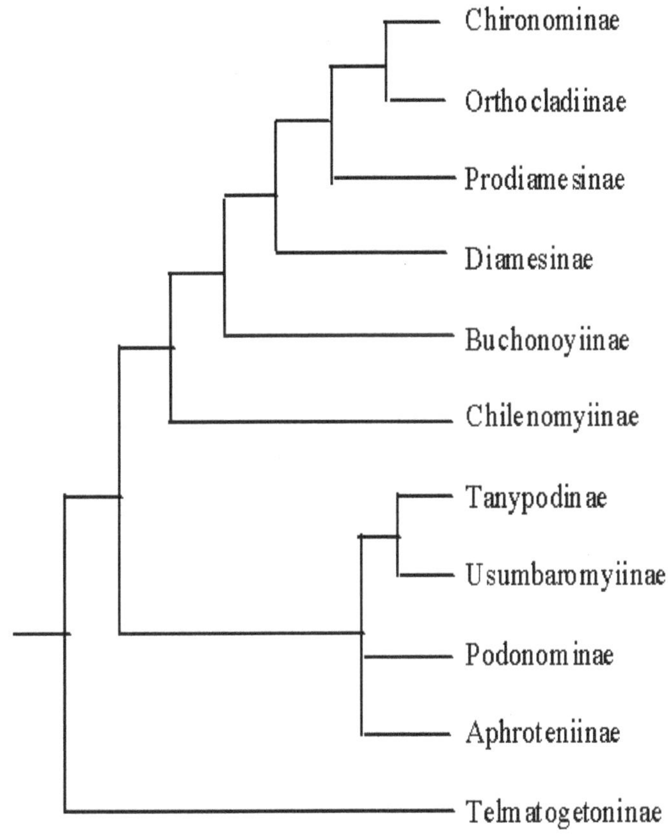


John Brackenbury (2000) observed about locomotory modes in the larva and pupa of *Chironomus plumosus*. The locomotory kinematics of *Chironomus plumosus* larvae and pupae were investigated in order to determine how different locomotory techniques may be related to possible underlying patterns of muscle activation and the particular life style and behavior of these juvenile stages. Larvae display three independent modes of motile activity: swimming, crawling and whole body respiratory undulation. Swimming and respiratory undulation involve the use of metachronal waves of the body bending which travel in a head to tail direction. Whereas swimming is produced by side to side flexures of the whole body, respiratory undulation employs a sinusoidal wave. Crawling appears to result from an independent program of muscle activation. Instead of a longitudinally transmitting metachronal wave of body flexure, a simultaneous arching of the body, combined with the alternating use of the abdominal and prothoracic pseudopods as anchorage points, produces a form of locomotion analogous to caterpillar looping. Larval swimming has a set speed and rhythm and is an 'all or nothing' locomotory maneuver but the neural program controlling larval crawling is adaptable; switching from a less to a more slippery substrate resulted in a shorter, faster stepping pattern. The pupa displays two swimming modes, somersaulting and eel like whole body undulation, the former being principally a brief, escape maneuver, the latter being a faster form of locomotion employed to deliver the pupa to the surface prior to adult emergence. Videography of larval and pupal movements was performed with a NAC 400 high speed video system (NAC Inc. Japan) with synchronized stroboscopic illumination generating 200 or 400 frames per second. Additional high resolution pictures were obtained using a Panasonic video camera with a normal frame rate of 25 per second. By using a stroboscope (Drelloscop 1018, Drelloscop Germany) at 50 Hz, it was possible to double the imaging rate to 50 fields per second, two separate images being obtained for each binary field normally making up a single frame. Data were recorded on a video cassette player equipped with a single field advance facility for detailed image measurement. Body profiles and swimming trajectories were traced by hand directly from the video screen. Experimental values are quoted as means ± 1 standard error. Paired values were compared for significant differences using the 't-test' at significance level of 5%. In total, approximately 50 larvae and 20 pupae were examined with an average of 3-4 measurements per parameter per animal.

Chaudhuri *et al.* (2001) prepared a checklist of Chironomid midges of the Indian subcontinent. It includes 313 species under 60 genera under four subfamilies. Four subfamilies are Diamesinae, Tanypodinae, Orthocladiinae and Chironominae. The types and paratypes of India and Bhutan are present in the National Zoological Collections (NCZ) at the Zoological Survey of India, Calcutta. They examined these types but unfortunately, most are represented only by fragments and were not helpful in determining the systematic positions of the species, therefore this checklist had been prepared on the basis of published works. An alphabetical list of 60 genera is as follows:

Subfamily- Tanypodinae

> Tribe- Pentaneurini

> > Genus- *Ablabesmyia* Johansen

> Genus- *Conchapelopia* Roback

> Genus- *Paramerina* Fittkau

> Genus- *Rheopelopia* Fittkau

Tribe- Procladiini

        Genus- *Djalmabatista* Fittkau

        Genus- *Procladius* Skuse

Tribe- Tanypodini

        Genus- *Tanypus* Meigen

Tribe- Coelotanypodini

        Genus- *Clinotanypus* Kieffer


Subfamily- Diamesinae

    Tribe- Boreoheptagyiini

        Genus- *Boreoheptagyia* Brundin

    Tribe- Diamesini

        Genus- *Diamesa* Meigen


Subfamily- Orthocladiinae

        Genus- *Allotrissocladius* Freeman

        Genus- *Brillio* Kieffer

        Genus- *Bryophaenocladius* Thienemann

        Genus- *Chaetocladius* Kieffer

        Genus- *Cricotopus* van der Wulp

        Genus- *Eukiefferiella* Thienemann

        Genus- *Heterotrissocladius* Sparck

        Genus- *Indocladius* Chaudhuri & Bhattacharyay

        Genus- *Limnophyes* Eaton

        Genus- *Metriocnemus* van der Wulp

        Genus- *Nasuticladius* Freeman

        Genus- *Orthoclaius* van der Wulp

        Genus- *Paracricotopus* Thienemann & Harnisch

        Genus- *Parakiefferiella* Thienemann

        Genus- *Parametriocnemus* Goetoghebuer

        Genus- *Paraphaenocladius* Thienemann

Genus- *Pseudoorthocladius* Goetghebuer

Genus- *Pseudosmittia* Goetghebuer

Genus- *Rheocricotopus* Thienemann & Harnisch

Genus- *Smittia* Holmgren

Subfamily- Chironominae

    Tribe- Chironomini

        Genus- *Beckidia* Saether

        Genus- *Chironomus* Meigen

        Genus- *Cladopelma* Kiefffer

        Genus- *Cryptochironomus* Kieffer

        Genus- *Demicryptochironomus* Lenz

        Genus- *Dicrotendipes* Kieffer

        Genus- *Einfeldia* Kieffer

        Genus- *Endochironomus* Kieffer

        Genus- *Gillotia* Kieffer

        Genus- *Glyptotendipes* Kieffer

        Genus- *Harnischia* Kieffer

        Genus- *Kiefferulus* Goetghebuer

        Genus- *Kribiocosmus* Kieffer

        Genus- *Lauterborniella* Thienemann &Bause

        Genus- *Microchironomus* Kieffer

        Genus- *Microtendipes* Kieffer

        Genus- *Nilodorum* Kieffer

        Genus- *Paracladopelma* Harnisch

        Genus- *Paratendipes* Kieffer

        Genus- *Polypendilum* Kieffer

        Genus- *Stenochironomus* Kieffer

        Genus- *Strictochironomus* Kieffer

        Genus- *Trichotendipes* Guha, Das, Chaudhuri & Choudhuri

        Genus- *Xenochironomus* Kieffer

Tribe- Tanytarsini

>	Genus- *Bernhardia* Datt &Chaudhuri
>
>	Genus- *Cladotanytarsus* Kieffer
>
>	Genus- *Microspectra* Kieffer
>
>	Genus- *Paratanytarsus* Thienemann & Bause
>
>	Genus- *Rheotanytarsus* Thienemann & Bause
>
>	Genus- *Tanytarsus* van der Wulp

Martinez *et al.* (2001) studied about heat shock regulatory elements present in telomeric repeats of *Chironomus thummi*. As in other Diptera, the telomeres of *Chironomus thummi* lack short telomerase specified repeats and instead contains complex sequences. They react to heat shock and other stress treatments by forming giant puffs at some chromosome termini, which are visible in polytene cells. All telomeres except the telocentric ends of chromosome 4L consist of large blocks of repeats, 176 bp in length. Three subfamilies of telomeric sequences have been found to show different distribution pattern between chromosome ends. TsA and TsC are characteristics of telomeres. Heat shock transcription regulatory elements have been identified in the telomeric sequences, appearing differentially represented in the three subfamilies, but otherwise rather similar in size and sequence. Interestingly TsA and TsB repeats share the well conserved heat shock elements (HSE) and GAGA motif, while the TATA box is only present in the former. Neither a HSE nor a TATA box appears in TsC repeats. Moreover experimental data indicate that the HSE is functionally active in binding heat shock transcription factor (HSF). These results provide for the first time, a molecular basis for the effect of heat shock on *Chironomus thummi* telomeres and might also explain the different behavior they show. A positive correlation between the presence of HSE and telomeric puffing and transcription under heat shock was demonstrated. This was also confirmed in the sibling species *Chironomus piger.*

Michailova *et al.* (2001) investigate the genotoxic effects of chromium on polytene chromosomes of *Chironomus riparius*. Chronic exposure to three different concentration of chromium (III) on polytene chromosomes of larvae of *Chironomus riparius* were tested from the embryonic stage to the fourth larval instar for two successive generations. In chromosome AB, CD, EF and G significant difference of chromosome aberrations were found between exposed and control larvae as well as changes in functional activity. Neither significant difference was found between the effects of the three treatments nor between the two generations. In chromosome G, the Balbiani ring system appeared as a model for studying the response of the genome to Cr (III) treatment. In approximately one third of the cells of the exposed larvae, the activity of the

Balbiani rings BRc and BRb was reversed. In 10% of the cells of both generations of treated larvae, deletion of collapse of BRc was observed. Pompons like G chromosomes were present in 6% of the cells. In 5% of the treated larvae the apical region of the chromosome G folded back so that so that the nucleolar organizer region appeared as if it was at the end of the chromosome. These structural and functional chromosomal changes are interpreted as a reaction of the genome to stressful rearing conditions. Chemical analysis of the Cr load in tissues of treated and control larvae were performed by atomic absorption spectrophotometry on a sample of 50 fourth instar larvae of each group. Perkin Elmer 500 spectrophotometer was used with a background correction with a deuterium lamp. Analysis was done performed according to Regoli & Orlando (1994). To identify chromosomal rearrangements, they used the standard chromosome maps of Hagele (1971) & Kiknadze *et al.* (1991). To check for deformities of the sclerotized mandibles, mentum, epipharyngeal pecten, premandible and antennae, and entomological preparation of the head capsule was performed for each larva. Value of Cr content in treated and control larvae were compared with the non parametric 'Curskal-Walker ANOVA' and frequencies of chromosomal aberrations in treated and control larvae were compared with G tests (Sokal & Rholf 1981).

Michailova *et al.* (2001) examined the effect of lead (Pb) nitrate on the structural and functional organization of the polytene chromosomes of *Chironomus riparius* larvae. No specimens with standard polytene chromosomes were found in the material treated with different concentrations of Pb ions. The polytene chromosomes of all individuals exposed to lead showed various somatic chromosomal rearrangements (heterozygous inversions, deletions, duplications and deficiencies), which were not detected in the studied controls. Deletions in chromosome G occurred in a high frequency resulting in the formation of so called "pompons". The activity of Balbiani rings (BRs) and nucleolar organizer (NOR) significantly decreased with increasing the concentrations of Pb ions. Sections of the polytene chromosomes where somatic aberrations were concentrated are considered as weak sites, it was shown that in these sites were distributed repetitive DNA elements (Alu and Hinf) that might be activated by stress agents and generated many chromosome rearrangements. Salivary polytene chromosomes of fourth instar larvae were used for cytogenetical analysis following the aceto-orcein method. Standard chromosome maps of Hagele (1970) & Kiknadze *et al.* (1991) were used to identify changes in the polytene chromosomes under stress agents. The location of repetitive DNA elements (Alu) was determined by means of the FISH techniques, following the protocol of Hankeln (1990). Alu probes were labeled by means of a Digissigenin kit. For morphological analysis, the appearance of the head capsule and body of the larvae was examined. The amount of lead associated with the substrate and Chironomids was measured by inductively coupled plasma optical emission spectroscopy (ICPOES) with a v-groove nebulizer and conventional Scott type spray chamber. Each treatment group was compared with each other and with control using a 'student t-test'. P<0.05 was considered as significant.

Guryev *et al.* (2001) described phylogeny of the Genus *Chironomus* inferred from DNA sequences of mitochondrial cytochrome b and cytochrome oxidase I. Two mitochondrial genes cytochrome b (*Cytb*) and cytochrome c oxidase subunit I (*COI*), have been used as phylogenetic markers in Chironomids. The nucleotide sequences of 685 bp from *Cytb* and 596 bp from *COI* have been determined for 36 *Chironomus* species from the Palaearctic, Holarctic and Australasia. The concatenated sequence of 1281 bp from both genes was used to investigate the phylogenetic relationships among these species. The nucleotide sequence alignments were used for construction of phylogenetic relationships among these species. The nucleotide sequence alignments were used for construction of phylogenetic trees based on maximum parsimony and neighbor joining methods. Both techniques produced similar phylogenies. Monophyly of the genus *Chironomus* is supported by a bootstrap value of 100% at the basal branch. Six clusters of species have been revealed with high bootstrap values supporting both monophyly of each cluster and the validity of the branching order within each cluster. Four species, *Chironomus circumdatus*, *Chironomus nepeanensis*, *Chironomus dorsalis* and *Chironomus crassiforceps* cannot be placed in any cluster. Cytological phylogenies were constructed using the same set of species except for *Chironomus biwaprimus*. These trees showed many similarities to that obtained from the mitochondrial sequence analysis, but also a number of significant differences. When compared with the tree constructed from the sequence of 23 species available for one of the globin genes, globin 2b (*gb2b*), there was better support for the mt tree than for the cytological trees. An intron which varies in its occurrence and position in *gb2b* was also investigated and the distribution of the introns supports the phylogenetic history of the genus *Chironomus* obtained with mt data. The differences observed in the cytological trees seem to be attributable more to the retention of the same chromosome banding sequence across several species, rather than convergent evolutionary events. An important question is the determination of the position of the subgenus *Camptochironomus* in relation to the representatives of the nominal subgenus *Chironomus* since it has been suggested that this is a separate genus. The *Camptochironomus* are internal to the trees and have arisen more recently than some of the species of the subgenus *Chironomus*, indicating that they are not sufficiently differentiated to be considered more than a subgenus.

Halpern *et al.* (2002) examined the protective nature of *Chironomus luridus* larval tubes against copper sulphate. A laboratory culture of an Israeli benthic midge, *Chironomus luridus* was exposed to copper sulphate. Two conditions were tested in bioassay experiments, one within silt tubes and other larvae without tubes. The non toxic, anionic, fluorescent dye, fluorescein was used to examine the effect of sublethal copper sulphate concentration on the permeability of cuticular, gill and gut epithelia of the Chironomids. Increased cell permeability which is the cause of cell damage was reflected by an increase in fluorescence intensity. Following exposure to copper sulphate, higher fluorescence was found in different body compartments: midgut, hindgut, tracheal gills, fat body, muscles and malpighian tubules. The effect was significantly higher in tube free larvae when compared to silt tube dwelling larvae. They concluded that in addition to its other functions in feeding, respiration and anti-predation shelter, the *Chironomus luridus* tube protects its inhabitant from toxins such as copper sulphate.

Pery *et al.* (2002) presented models to link feeding with growth, emergence and reproduction of the *Chironomus riparius*. Models were based on three assumptions. First assumption was that the length to width ratio remains constant during growth. This assumption was termed isomorphism. Second assumption was that maintaining energetic costs like losses due to respiration are much lower than growth during the larval instars. Third assumption was that a maximum length exists can be different for males and females. In their experiments they found that a) Mean daily weight increase as a function of daily feeding input. b) Length growth pattern with food *ad libitum* as a function of time increases. c) Length growth pattern for two diets and three starting densities is positively correlated. d) There is positive correlation between diet and numbers of eggs given by female.

Hapern *et al.* (2003) reported non-O1 and non-O139 *Vibrio choerae* from Chironomid egg masses from different fresh water bodies in Israel, India and Africa. The bacterial population increases as water temperature surpassed 25°C. Thirty five different serogroups of *Vibrio cholerae* were identified among the bacteria isolated from Chironomids, demonstrating population heterogeneity. Two strains of *Vibrio cholera*, O37 and O201 that were isolated from Chironomid egg masses in Zanzibar Island were NAG-ST positive. Above findings support the hypothesis that the association found between Chironomids and the cholera bacteria is not a rare coincidence, indicating that Chironomid egg masses may serve as yet another potential reservoir for *Vibrio cholerae*.

Halpern *et al.* (2003) reported that *Vibrio cholerae* O9, O1 and O139 supernatants lyses the gelatinous matrix of the Chironomid egg mass and inhibit eggs from hatching. The extracellular factor responsible for the degradation of Chironomid egg masses was purified from *Vibrio cholerae* O9 and O139 and was identified as the major secreted hemaglutinin/protease (HA/P) of *Vibrio cholerae*. The substrate in the egg mass was characterized as a glycoprotein. These findings showed that HA/P plays an important role in the interaction of *Vibrio cholerae* and Chironomid egg masses.

Henriques *et al.* (2003) observed the feeding behavior of the Chironomid larvae present in the Rio-da Fazenda, (Rio-de Janeiro, Brazil). Algae, fungi, pollen, leaf and wood fragments, animal remains, detritus and silt are the main gut contents found in the larvae studied. The main food item ingested by the larvae was detritus except for the *Stenochironomus*, whose main food source was leaf and wood fragments. Tanypodinae exhibited a large quantity of animal remains of several kinds in the diet. During the period studied it was observed that the diet of 16 genera (out of 24 studied) varied. Tanypodinae had mainly coarse particulate organic matter (>1mm) in the gut contents, while Chironominae and Orthocladiinae had fine particulate organic matter (<1mm).

Nazarova *et al.* (2004) observed the buccal deformities in Chironomid larvae from Santa Marta, Columbia. The mouthpart deformities of Chironomid larvae are considered as indicators of environmental stress caused by water pollution such as heavy metals, pesticides and organic contamination. The most frequently met abnormalities in genus *Goeldichironomus* was abrasion and breakage of the mentum and/or mandibles, which reached up to 96.5% of all specimens of *Goeldichironomus carus*. In all other taxa such specific morphological responses were not found. There the amount of abnormalities was relatively low. In most cases the abnormalities concerned lacking or misshapen teeth owing to high abrasion. Abnormalities observed in Tanypodinae were apparent on the hypophayngeal structures.

Michailova *et al.* (2005) analyzed the chromosome variability in three *Chironomus* species *viz.*: *Chironomus plumosus*, *Chironomus muratensis* and *Chironomus annularis* from the natural populations in Poland. A comparative analysis of band sequences with other Palearctic populations was done. For this experiment, larvae were fixed in a solution of 96% ethanol and glacial acetic acid (3:1). Cytogenetic and morphological preparations were obtained from each larva. Isolated salivary glands were squashed for polytene chromosome using a aceto-orcein method. Freezing in liquid nitrogen after removal of cover slips did permanent slides. After several steps in alcohol and xylene, they were remounted in Euparal for morphological analysis. Slides of polytene chromosomes and larvae were deposited at the Institute of Zoology, Sofia-Bulgarian Academy of Sciences. Polytene chromosomes of *Chironomus plumosus* were identified by applying the maps of butler *et al.* (1999), Gunderina *et al.* (1999), Golygina and Kiknadze (2001). The polytene chromosomes of *Chironomus muratensis* were identified by comparing with the literature (Kiknadze *et al.* 1991; Michailova *et al.* 2002). For a subsequent detailed karyotype analysis of *Chironomus annularis*, standard chromosome maps done by Keyl & Keyl (1959), Keyl (1962), Petrova and Michailova (1986) and Kiknadze *et al.* (1991, 1996) were used. Inverted and standard band sequences were considered as alleles and their frequencies were estimated according to Butler *et al.* (1999). A dendrogram was made using

UPGMA cluster analysis (Sneath and Sokal 1973) with PHYLIP software. Heterozygous frequencies of *Chironomus plumosus* were tested for confirming to expectation under Hardy-Weinberg equilibrium using $^2$ test. Chromosome band sequence of *Chironomus plumosus* and *Chironomus muratensis* did not differ from other Palearctic populations. A new homozygous band sequence was discovered in *Chironomus annularis*.

Habashy (2005) investigated the rearing of Chironomid larvae under laboratory conditions to study the growth rate (weight and length) of Chironomid larvae under different feeding systems. Three types of food were used, the first was Tetramin flaked commercially available fish food (D1), the second food was algae (*Scenedesmus* sp., D2) and the 3rd was baker yeast (D3). Larvae had mean initial length 1.5mm and mean weight 0.99 µg. Three replicates were performed for each treatment. The three diets had significantly ($P<0.05$) effect on both weight and length of Chironomid larvae, but larvae fed on D1 showed the highest weight and length which represented by 36.68 µg and 9.29 mm respectively. Followed by those fed on D3 was represented by 7.03 mm for length and 31.03 µg for weight, while D2 showed the lowest value for both length (6.28mm) and weight (27.08 µg). The highest percentage of protein (51.15%) was for larvae fed on D1 followed by those fed D3 (49.45%), while the larvae fed on D2 showed the lowest value of protein (26.45%).

Ebrahimnezhad and Fakhri (2005) did the taxonomic study of Chironomidae larvae of Zayandehrood River, Iran. They also studied the effects of selected ecological factors on their abundance and distribution. They did the comprehensive study on the Chironomid larval identification in running waters of the country particular in large rivers. Samples were collected from 9 sites and were hand sorted in the laboratory and the larvae were identified to the generic level using available identification keys. To study the effects of sites and seasons and selected ecological factors on Chironomid larval abundance and distribution, data were analyzed using a two way ANOVA.

Ozkan and Camur (2006) studied the dynamics of Chironomid larvae and some physicochemical features of water in Meric River (Turkey). According to Shannon-Weiner diversity index, Meric River has diversity 1.23. Furthermore according to Pearson correlation index, water temperature ($r=+0.71$, $P<0.05$), pH ($r=+0.61$, $P<0.05$) and chloride ($r=+0.61$, $P<0.05$) had direct proportional while $NO_2/N$ ($r=-0.73$, $P<0.05$) had inverse proportion with number of larvae.

Lee *et al.* (2006) evaluated pollutant induced expression of heat shock proteins (HSPs) and hemoglobin genes in the larvae of the *Chironomus tentans* to identify a sensitive biomarker of freshwater monitoring. The response of the HSPs gene expression by chemical exposure was rapid and sensitive to low chemical concentrations but it was not stress specific. An increase in the expression of HSPs genes was observed not only in a stress inducible form

(HSP70) but also in a constitutively expressed form (HSC70). The expression of Hemoglobin genes showed chemical specific responses: that is alkyl phenolic compounds increased the expression of hemoglobin genes, whereas pesticides decreased expression. Statistical differences between the control and treated larvae were examined with the aid of parametric t-test using SPSS 12.0 KO for all analysis.

Cranston (2006) reported the *Xylochironomus kakadu*, a new genus and species of wood mining Chironomid from northern Australia. Formerly known by a code 'unknown genus K1', the larvae mine soft immersed timber in tropical Australia. All type material is mounted on slides in Euparal and deposited in the Australian National Insect Collection, CSIRO Entomology, Canberra, Australia.

Al-Saffar (2007) observed the larval mouthpart deformities in *Chironomus annularis* from Al-Hammar Marsh, Southern Iraq and Tanjero River, Kurdistan and Northern Iraq. He showed that deformities in the pectens, menta and mandibles can be sign of low, medium and high levels of contamination respectively. The deformed pecten epipharyngis is possibly a result of either DDT or heavy metals.

Pramual *et al.* (2008) studied the population cytogenetics of *Chironomus circumdatus* Kieffer, 1921 from Thailand. *Chironomus* larvae were collected from 24 populations in six provinces from northeastern Thailand during August to January. Larvae were fixed in carnoy's fixative (3:1, 95% ethanol:glacial acetic acid) and stored at 4°C. Salivary gland polytene chromosomes were prepared using the aceto-orcein method. Polytene chromosome banding patterns were read band for band in comparison with the standard polytene chromosome map and chromosome inversions were recorded. Population genetic structure was analyzed by means of Wright's F-statistics including $F_{IS}$, the measurements of the heterozygous deficit within population, $F_{ST}$ is among populations and $F_{LT}$ the deficit of the heterozygous overall population. The F-statistics were calculated following the method of Weir and Cockerham (1984). All of the above analyses were performed using the software FSTAT (Groudet 1995). Nei's modified genetic distance $(D_A)$ based on inversion frequencies were calculated to assess levels of genetic differentiation between populations using the software DISPAN Ota 1993). A Mantel (1967) test was used to determine the relationship between genetic distance $(D_A)$ and geographic distance (log Km). An UPGMA population tree based on $D_A$ was calculated using MEGA 3.1 (Kumar *et al.* 2004). The relationship between inversion frequency and environmental condition was examined using the Pearson correlation. A total of 1505 larvae from 24 populations were examined cytologically. 12 chromosomal inversions were found and most of these (9 of 12) were rare inversions. All populations were in Hardy-Weinberg equilibrium. Population genetic structure analysis indicated significant genetic differentiation between populations ($F_{ST=}$

0.037, P<0.001). Geographic distance was the principal factor limiting gene flow between populations. Nei's modified genetic distance ($D_A$) between populations ranged from 0.001 to 0.011 with an average of 0.003. An UPGMA population phenogram depicting relationship between populations based on $D_A$ values revealed three groups of populations group I, II and III each characterized by different inversions/inversion frequencies. Significant correlation of inversion might have a role to play an adaptation to high temperature habitat.

Pfenninger and Nowak (2008) observed reproductive isolation and ecological niche partition among larvae of the morphologically cryptic sister species *Chironomus riparius* and *Chironomus piger*. They showed that these morphologically cryptic sister species partition their niches due to a certain degree of ecological distinctness and total reproductive isolation in the field.

Silva *et al.* (2008) analyzed the functional feeding habits of Chironomidae larvae in a lotic system from mid-western region of Sao Paulo State, Brazil. Collectors were the dominant organisms, represented by genera *Chironomus*, *Fissimentum* and *Cryptochironomus* following shredders and predators.

Sanseverino and Nessimian (2008) studied about the food of larval Chironomidae in submerged litter in a forest stream of the Atlantic forest (Rio de Janerio, Brazil). In order to study feeding habits they investigated the gut contents of 23 genera of Chironomidae and Orthocladiinae larvae. Algae, fungi, pollen, particulate organic matter, silt and plant fibers were observed in the larval guts. Particulate organic matter was the predominant food source and according to chi-square distribution test, the proportion of food items consumed by the larvae showed seasonal variation except for *Stenochironomus*. The largest size (280μm) of ingested food particle was recorded for *Xestochironomus* and the smallest (10μm) for *Lepiscladius*. A cluster analysis dendrogram showed that the groups of genera were separated according to season on which the smallest size of food particle was observed. Dendrogram was produced by UPGMA (unweighted pair-group method with arithmetic averages) clustering of this resultant matrix with NTSYS- PC 1.7 (Rohif, 1992).

Henry and Santos (2008) investigated about the importance of excretion by *Chironomus* larvae on the internal loads of nitrogen and phosphorus in a small eutrophic urban reservoir. Ammonium and phosphate excretion rates by *Chironomus* larvae of small size (6-10mm) were significantly higher than those of the Chironomids having medium (9-11mm) and large (11-16mm) sizes. Through a linear relation between biomass (dry weight) and total length and between excretion and biomass and  data on Chironomid densities, after an intense sampling in 33 sites distributed all along the reservoir bottom, the mean phosphate and ammonium excretion rates corresponded to 2014 ± 5134 μg.m$^{-2}$/day and 1643±3974 μg.m$^{-2}$/day respectively. Considering the mean biomass (34mg.m$^{-2}$) of *Chironomus*, the lake area (88156 m$^2$) and the mean excretion rates, the contribution of benthic Chironomids to the internal loads would be 181 kg P and 147 kg N. These values showed that the internal loads by excretion from *Chironomus* larvae correspond to approximately 33% of the external loads of phosphorus in the lake and in the case of nitrogen to only 5%.

Petrova and Zhirov (2008) described the polytene chromosomes of Chironomid larvae collected on the Wrangel Island (Russia). Larvae have 2n=8: AB CD EF and G (*Chironomus thummi* Keyl, 1962 complex). By karyotype larvae are identical to *Chironomus* sp. Le1, described from the Sagystyr Island of the Vst-Lena reserve (Yakutia) (Kiknadze *et al.*, 1996). The insignificant differences include the number of nucleoli and puffs, which was regarded as result of habituation of populations under different ecological conditions. For this experiment material was fixed in the mixture of 96% ethanol and glacial acetic acid (3:1). The preparations were made using the aceto-orcein technique (Chubareva, Petrova, 1982).

Mi-Hee Ha and Jinhee Choi (2008) studied the effects of environmental contaminants on hemoglobin in the 4th instar larvae of *Chironomus riparius* with respect to the total hemoglobin contents, individual globin gene expression, individual globin protein expression and hemoglobin oxidation. The result suggested that globin can be a target molecule of environmental contaminants and of the tested parameters; the alteration of individual globin levels (i.e. mRNA and protein levels) may have potential for the development of a biomarker for ecotoxicity monitoring.

Kiknadze and Istomina (2010) studied the karyotype structure and chromosomal polymorphism in several European and Asian populations of *Chironomus luridus* Strenzke, 1959. *Chironomus luridus* larvae of fourth instar were used in the work. The larvae were fixed in a mixture of 96% ethanol and glacial acetic acid (3:1) and stored in a refrigerator. Squash preparations of the salivary gland polytene chromosomes were made using aceto-orcein method (Kiknadze *et al.*, 1991). Polytene chromosome arms A, E and F were mapped according to Keyl (1962) and arms C and D, according to Devai *et al.* (1989), using the banding sequences of *Chironomus piger* Strenzke, 1959 polytene chromosomes as a standard. Arms B and G were not mapped due to complex rearrangements as compared with *Chironomus piger*. Inversion banding sequences of polytene chromosomes were designated using the abbreviated species name, arm designation and banding sequence number (lur A1, lur A2 etc.). The studied cytological slides and fixed larval body after dissection of salivary glands are preserved in the collection of the Institute of Cytology and Genetics of Russian Academy of Sciences, Novosibirsk, Russia. Axioskop 2 plus microscope, CCD camera Axiocam HRc, software package Axiovision 4 (Zeiss, Germany) were used in this work. Inversion polymorphism was detected in six of the seven chromosome arms: three banding sequences detected in arm A, six sequences in arm B, and two sequences in arm G. Only arm D was monomorphic in all studied populations. In total 22 banding sequences were recorded in *Chironomus luridus*. The Asian populations were less polymorphic containing only 8 to 10 sequences, whereas the European populations had 11 to 16 sequences. The new banding sequences lur A3, lur B5, lur  E3 were found in Asian populations, whereas the sequences lur B2, lur B3, lur B4, lur B6, lur  E2, lur F1, lur F2a, lur F3 lur F4 and lur G2 were lacked. The total level of inversion heterozygosity in the Asian populations was 12-25% versus 78-80% in the European populations.

Ozkan *et al.* (2010) did the ecological analysis of Chironomid larvae in Ergene River basin (Turkish Thrace) in their study. They described the physicochemical properties of the water to analyze the relationships between species composition of Chironomid communities and environmental parameters. Larval Chironomid samples were collected using Ekman Grab (15cm*15cm) and then washed through 0.5mm sieve and preserved in 70% ethanol prior to identification. Chernovskii (1961), Fittkau (1972), Beck & Beck (1969), Bryce & Hobart (1972), Saether (1977), Fittkau & Reiss (1978), Moller-Pillot (1978-1979; 1984), Fittkau & Roback (1983), Sahin (1984,1987,1991), Sahin *et al.* (1988). Armitage *et al.* (1995) and Epler (2001) work was used to identify the larvae at the lowest possible taxonomic level. Evaluation of species abundance (number of individuals/ $m^2$) and value of specific richness enabled them to determine the Shannon's index. Some physicochemical parameters including water temperature, pH and conductivity were measured (using ordinary thermometer, Jenway 3040 mark ion analyser and WPA CM35 mark conductivity meter, respectively) at the time of sampling of benthos. Water samples taken by Ruttner water sampler were carried to the laboratory to measure the other parameters including dissolved oxygen (using wrinkler method), BOD and COD, $Ca^{++}$, $Mg^{++}$, $Cl^{-1}$, $NO^2$, $NO^3$ and $PO_4^{-3}$ using classical titrimetric and spectrophometric methods (Egmen & Sunlu, 1999). Quality grades of the water at the sampling sites were determined using National water quality standards (SKKY, 2004). Correlation between data of both larval number and physicochemical parameters were determined statistically by using Spearmann correlation index in SPS 9.0 for Windows. Furthermore the similarities for larval composition were determined using Bray-Curtis index in terms of both sampling sites and sediment types. It was found that sediment nutrient enrichment strongly influenced Chironomid assemblage structure. According to Shannon index, species diversity of the basin was found as H`=1.41 at average. Spearman index had indicated that significant correlations were found between Chironomid larval densities and some environmental variables such as conductivity (r=0.708), dissolved oxygen (r=+0.810), BOD (r=+0.822) and COD (r=+0.805) for P<0.05.

Kiknadze *et al.* (2010) described the karyotype of *Chironomus uliginosus* Keyl. Four *Chironomus uliginosus* populations from the Netherlands and one population from Germany were investigated. Fourth instar larvae were fixed in 3:1 mixture of 100% ethanol and glacial acetic acid for cytogenetic study. Isolated salivary glands were squashed for polytene chromosome preparation. Polytene chromosome squashes were made by the aceto-orcein method (Kiknadze *et al.* 1991). The polytene chromosomes were mapped in accordance with Keyl's system (1962) for arms A, E, F and Devai *et al.*'s system (1989) for arms B, C, D. Banding sequences of all seven chromosome arms in the *Chironomus uliginosus* karyotype were analyzed. Inversion polymorphism was observed in arms A, B, C, D, E and F. Total of 18 banding sequences were recorded. The sets and number of banding sequences varied between populations causing their cytogenetic differentiation. The comparison of banding sequence sets and frequencies between the Netherlands and Russian populations of *Chironomus uliginosus* was carried out and considerable cytogenetic differences between these distant populations were observed.

Al-Shami *et al.* (2010) studied the morphological deformities in parts of the head capsule of *Chironomus* spp. larvae inhabiting three polluted rivers (Permatang Rawa, Pasir and Kilang Ubi) in the Juru River basin, northeastern peninsular Malaysia. Samples of the fourth instar larvae in each River were collected and examined for deformities of the mentum, antenna, mandible and epipharyngis. The total deformities correlated closely with deformities of mentum but only weakly with deformities in other parts of head. The total deformity incidence was strongly correlated with high contents of sediment Mn and Ni. The mentum and epipharyngis deformities incidence was highly correlated with an increase of TSS (total suspended solids), total aluminum and ammonium and a decrease in pH and dissolved oxygen.

Morais *et al.* (2010) studied the diversity of larvae of littoral Chironomidae and their role as bioindicators in urban reservoirs of different trophic levels located in hydrographic basin of the Paraopeba River, an affluent of the Sao Francisco River basin; the Serra Azul reservoir, Vargem das flores reservoir and Ibirite reservoir (Brazil). The reservoirs were sampled every three months during the dry season (June and September) and during the wet season (March and December). Along the littoral zone of each reservoir, 30 samples were collected using a Eckman-Birge ($0.0225m^2$) sampler. The samples were deposited in plastic bags and transported to the laboratory, where they were washed on sieves of 1 mm and 0.5 mm meshes (Larsen *et al.*, 1991). Sub surface water samples were collected using Van Dorn bottles, for the measurement of the physical and chemical parameters. The physical and chemical parameters of the surface water (pH, temperature, dissolved oxygen, electrical conductivity and turbidity) were measured in situ, using a multi analyzer and portable apparatus (YSI). The secchi disc was used to evaluate the depth of the trophic zone. For the measurement of total nitrogen, total phosphorus and orthophosphates, 30 water samples were collected from each reservoir and transported to the laboratory in refrigerated polyethylene bottles. These measurements were performed according to the standard methods for the examination of water and waste water (APHA, 1992). Chironomid larvae were treated with a 10% lactophenol solution and identified under a microscope (400x) with the aid of taxonomic keys (Trivinho & Strixino 1995; Epler 2001). The occurrence of morphological deformities in the mentum was recorded and counted. All Chironomids collected were analyzed. The lack or excess of teeth, asymmetry, fusion, tooth malformation and combinations of these characteristics were considered deformities. The Shannon diversity index, Pielou's equitability index (Magurran 1988), organism density (individuals/m$^2$) and taxonomic richness (total number of taxa in each sample) were calculated in order to evaluate the structure of the Chironomid assemblages. An ANOVA (software STATISTICA for Windows 5.1) of the data on the composition of the Chironomid assemblages was used to evaluate if there were significant differences among the three reservoirs. A cluster analysis (software Primer 6 beta 2004) was performed in order to assess the similarity in the taxonomic composition of the assemblages found in the three reservoirs. The Bray-Curtis index and UPGMA (unweighted pair group method with arithmetic mean) were used as the amalgamation method. Using the three reservoirs as the groups, an indicator species analysis (Dufrene and Legendre 1997) using the PC-Ord software (version 3.11, 1997) was carried out in order to establish the indicator taxa for each reservoir. The potential indicator taxa for each reservoir were established through an indicator species analysis. *Fissimentum* was the indicator taxon in Serra Azul, whereas *Pelomus* was the indicator taxon in Vargesm das flores and *Chironomus* in Ibirite.

Lencioni *et al.* (2011) studied the diversity and distribution of Chironomids in pristine Alpine and pre-Alpine springs (Northern Italy). A total of 173 macro invertebrate samples were collected in which 26,871 Chironomids (including larvae, pupae, pupal exuviae and adults) were counted. Five subfamilies (Tanypodinae, Diamesinae, Prodiamesinae, Orthocladiinae and Chironominae), 54 genera and 104 species/groups of species were identified. As expected, Orthocladiinae accounted for a large part of specimens (82%) followed by Diamesinae (10%), Tanytarsini (6%) and Tanypodinae (2%). Together the Chironominae and Prodiamesinae contributed less than 0.05% of the fauna. Larvae represented 97.5% of specimens, mostly occurred at intermediate altitude (900-1200 a.s.l.).

Bhaduri *et al.* (2011) studied about response of the Chironomid larvae to the environmental condition. They correlated heavy metal pollution in aquatic bodies and incidence of chromosome aberration in *Chironomus striapennis* Kieffer. The chromosomes observed in the cells of the samples collected from the polluted water bodies showed presence of polymorphic polytene chromosomes.

Michailova *et al.* (2011) evaluated the effects of pollution on the biodiversity and genome response of Chironomid larvae in the trace metal contaminated water environment. No change on the Chironomid species diversity was found. The higher concentration of trace metals (Cd, Pb, Cr and Zn) affect the genome of 5 cytogenetically studied Chironomid species: *Chironomus bernesis, Chironomus plumosus, Chironomus* sp. 1, *Kieffferulus tendipediformes* and *Glyptotendipes cauliginellus*. Genome instability of Chironomid larvae was manifested by two ways, one was fixed chromosome rearrangements and another was somatic structural and functional alterations in salivary gland chromosomes of cytogenetically studied Chironomidae species.

Wulker *et al.* (2011) investigated the karyotypes of six African *Chironomus* species (*Chironomus alluaudi* Kieffer, *Chironomus transvaalensis* Kiefffer, *Chironomus* sp. Nakuru, *Chironomus formosipennis* Kieffer, *Chironomus* prope *pulcher* Wiedemann and *Chironomus* sp. Kisumu). Chironomid larvae and pupae were collected with the aid of pond net. Extra samples of larvae, pupae, pupal exuviae and adults were taken using tweezers, drift and sweep nets. Samples were preserved in 75% ethanol. Chironomids were mounted on slides and identified to species/groups of species according to Serra-Tosio (1971), Pinder (1978), Ferrarese & Rossaro (1981), Rossaro (1982), Nocentini (1985), Schmid (1993), Jenecek (1998), Stur & Ekrem (2006), Lencioni *et al.* (2007) and Rossaro *et al.* (2009). During each biological survey environmental variables were recorded. Altitude was measured by GPS with an instrument with error of about 10-15 metre. The percent grain size composition of the substratum was evaluated visually as percentage of gravel, cobbles, sand, silt, stones and rocks. Water samples were collected in acid cleaned graduated bottles for hydro-chemical analysis (conductivity, alkalinity, hardness, dissolved oxygen, % oxygen, saturation, pH, nutrients, anions, cations and metals). Analysis was performed using standard methods following the American Public Health Association (APHA, AWWA & WEF 2005). Water temperature was measured with a field multiprobe (Hydrolab). In addition, the canopy cover was visually estimated in five classes: 0%, 25%, 50%, 75% and 100%. Discharge was measured using a graduated bucket. Average current velocity was measured with an OTT propeller-flow meter. Turbidity was recorded using a portable turbidimeter MicroTPI. Shannon-diversity index was calculated for each spring with the MVSP 3.1 computer package. Pearson correlation between biological and environmental variables was calculated with STATISTICA 8.0 computer package. Values with P<0.05 were considered significant. Ordination analysis was carried out by means of the CANOCA 4.5 computer package. A k-means cluster analysis was carried out to cluster sites into similar groups based on Chironomid taxon assemblages. A discriminant analysis based on the Wilk's lambda test was performed to detect the environmental factors separating the 5 k-means groups. Values with P<0.05 were considered significant. Of the six

*Chironomus* karyotypes, three had "pseudothummi" cytocomplex chromosome arms combinations AE CD BF and G (*Chironomus alluaudi*, *Chironomus transvaalennis* and *Chironomus* sp. Nakuru), two had "thummi" cytocomplex arm combinations AB CD EF and G (*Chironomus formosipennis* and *Chironomus* prope *pulcher*) and one had "parathummi" arm combinations AC BF DE and G (*Chironomus* sp. Kisumu).

Michailova *et al.* (2012) showed that Chironomids and their salivary gland chromosomes can serve as indicators of trace metal genotoxicity. These larvae exhibit aberrations in their polytene chromosomes on exposure to trace metals (Cr, Al, Pb and Cu). The genome response consisted in a statistically significant increase in somatic chromosome aberrations and decrease in BR and NOR activity to levels lower than those of larvae under standard conditions. The main chromosomal aberrations are inversions, amplifications, deletions and deficiencies. Deletions in the chromosome G in *Chironomus riparius* transformed it into a pompon like structure.

Eggermont and Heiri (2012) reviewed the nature of relationship between Chironomids and temperature based on the available ecological evidence. After discussing many of the surveys describing the distribution of Chironomid taxa in lake surface sediments in relation to temperature, they examined evidence from laboratory and field studies exploring the effects of temperature on Chironomid physiology, life cycle and behavior. These findings suggested that indirect effects of temperature on physical and chemical characteristics of lakes play an important role in determining the distribution of lake living Chironomid larvae. However they also demonstrated that no single indirect mechanism has been identified that can explain the strong relationship between Chironomid distribution and temperature in all regions and database presently available.

Marinkovik *et al.* (2012) studied the response of the *Chironomus riparius* to mutigeneration toxicant exposure. For this they performed a multigeneration experiment in which they exposed *Chironomus riparius* laboratory cultures for nine consecutive generations to two exposure scenarios of respectively copper, cadmium and tributylin. Total emergence and mean emergence time were monitored each generation, while the sensitivity of the cultures was assessed at least every third generation using acute toxicity tests. They observed that the sublethally exposed cultures were exposed to substantially higher toxicant concentrations after the sixth generations were severely affected in the eighth generation followed by signs of recovery. They concluded that *Chironomus riparius* can indeed withstand long term sublethal toxicant exposure through phenotypic plasticity without genetic adaptations.

Majumdar and Gupta (2012) observed the effects of acute and chronic toxicity of copper on *Chironomus ramosus* from Assam, India. Acute toxicity of copper on

*Chironomus ramosus* was determined by exposing third instar larvae to graded concentrations of copper sulphate ($CuSo_4.5H_2O$). Median lethal concentrations ($LC_{50}$) of Cu as $CuSo_4$ at 24, 48, 72 and 96 hr were determined as 3280, 1073.33, 780 and 183 $\mu gl^{-1}$ respectively. For determining the effects of chronic toxicity, small first instar larvae were individually exposed to sublethal concentrations of $CuSo_4$ (1-8 $\mu gl^{-1}$) for a period of 21 days. Discoloration and thinning of body were detected at $1\mu gl^{-1}$ and ventilation movements, pupation and adult emergence were significantly affected at 1.8 $\mu gl^{-1}$. At 10 $\mu gl^{-1}$ concentration, growth and tube building activities of the larvae were significantly different from the control.

Midya *et al.* (2012) studied the polytene chromosomes of *Chironomus striapennis* Kieffer in terms of pollution in natural habitats. The polytene chromosomes obtained from the flies collected from the polluted environment, exhibited asynapsis along the polytene chromosome arms. Asynapsis along the fourth polytene chromosome in this species was most prominent. The larvae grown in artificially developed polluted condition in the laboratory with addition of Cu as heavy metals in the culture medium also exhibited asynapsis of the fourth polytene chromosome and at higher dose *i.e.* 15 mg/kg of soil in the substratum of the culture tray, the larvae showed almost complete asynapsis of the fourth chromosome.

Petrova *et al.* (2012) studied the cytotaxonomy and morphology of Chironomid larvae in Armenia. In their study several species of *Cricotopus* genus of subfamily Orthocladiinae, *Diamesa* genus of subfamily Diamesinae and *Chironomus* genus of subfamily Chironominae had been identified. Larvae were collected by net with the ring of 15 cm in diameter. The collected material was fixed at collection site in Carnoy's fixative (3:1 ethanol 96% : ice acetic acid). To make the karyological preparations salivary glands were isolated in a drop of lactic acid, stained for 20 minutes with 2% aceto-orcein solution and after the second maceration in 45-60% lactic acid, cells were separated.

Midya *et al.* (2012) studied the effect of lead as heavy metal in aquatic habitat to promote polymorphism of polytene chromosomes on *Chironomus striatipennis* Kieffer. The larvae of the species hatched out from the eggs were cultured in the media containing different doses at lead nitrate such as 5, 10, 15  and 20 mg of Pb $(NO_3)_2$ mixed with soil in the substratum of culture trays. The larvae were allowed to grow in such Pb mixed culture media for a period so as to get the penultimate fourth instar larva. Polytene chromosomes prepared from the fully grown larvae showed appearance of polymorphic polytene chromosomes in their salivary gland cells. Asynapsis, heterozygous deletions and inversions were the causes behind occurrence of polymorphism. This indicated that Pb in the culture media develops unhealthy condition with which the larvae combat by rearranging the genomic constitution and such rearrangements have been manifested through the appearance of aberrant polytene chromosomes.

Ebau *et al.* (2012) investigated the acute toxicity of cadmium and lead on larvae of two tropical Chironomid species, *Chironomus kiiensis* Tokunga and *Chironomus javanus* Kieffer. They concluded that cadmium was more toxic to the Chironomids than lead and *Chironomus javanus* was significantly more sensitive to both metals than *Chironomus kiiensis* (P<0.05).

Xin Qi *et al.* (2012) described a new species of the genus *Microtendipes* Kieffer, Microtendipes *zhejiangensis* sp. n. from oriental China. They also provided the key to the males of this species. The material examined was mounted on slides following the procedure of Saether (1969). Specimens are deposited in the College of Life Sciences, Nankai University, China and College of Life Sciences, Taizhou University, China.

Ragonha *et al.* (2013) observed the influence of shoreline availability on the density and richness of Chironomid larvae in Neotropical floodplain lakes of upper Parana River. The shoreline development index was created for temperate lakes. However Neotropical lakes have different geological formations, so they proposed a modified index. They found that there was a strong positive relationship between the shoreline development index and Chironomid density (R2=0.37; P<0.001).The regression model to predict Chironomid density is as follows-

Density of Chironomidae = 3.084+0.506*DL

They also found positive relationship between the shoreline development index and Chironomid richness (R2=0.45; P<0.001). The regression model to predict Chironomid richness is as follows-

Richness of Chironomidae = 9.472+9.619*DL

Rico and Quesada (2013) described the ecology and distribution of two native Antarctic Chironomid species *Parochlus steinenii* and *Belgica antarctica,* both found on Byers Peninsula (Livingston Island, South Shetland Islands). Chironomid samples in streams and lakes throughout Byers Peninsula were collected during several expeditions either with a hand net (200 μm mesh size) by Kick or sweep sampling or with a Surber sampler (200 μm mesh size and 30*30 cm surface). Chironomids were fixed in the field using 4% formaldehyde and subsequently preserved in 70% ethanol. Some samples were processed *in vivo* in the laboratory where the animals were either preserved directly in absolute alcohol or frozen for isotopic analysis. Isotopic $^{13}$C analysis was made using an Elemental Analyzer Carlo Erba 1108 coupled to an isotope ratio mass spectrometer micromass isochrome in continuous flow mode. Isotopic studies show a non selective feeding regime for both species with mixed carbon sources associated with both biofilm/microbial mats. *Parochlus steinenii* inhabits lakes of the central plateau of Byers Peninsula associated with aquatic mosses on the bottom of lakes and in some streams of the south beach area. Some streams have stable populations which are able to complete their life cycle while other streams have temporary unstable populations. *Belgica antarctica* also inhabits streams running through mosses located in the south beach area. This species has a limited dispersed capacity. Both species coexists on Byers Peninsula and share some stretches of streams.

Vedamanikam *et al.* (2013) observed the effects of four heavy metals-copper, cadmium, mercury and zinc on five species of *Chironomus* larvae *viz.*: *Clunio gerlachi*; *Paramerina minima*; *Gymnometriocnemus mahensis*; *Tanypus complanatus* and *Larsia pallidissima* present on Mahe Island of Seychelles. These species are endemic to the Island. They were exposed to four heavy metals for a 96-hr period and the $LC_{50}$ values were calculated. The metal salts selected for this investigation were zinc chloride, mercuric chloride, copper (II) chloride and cadmium (II) chloride. All metal salts were of analytical grade supplied by MERCK, Germany. One gram per liter stock solutions were prepared for each metal. The range finding test consisted of a series of seven concentrations that different by a factor of 10%. All tests were conducted with three replicates and with controls. Test containers with volumes of 600 ml were selected for execution of the bioassays. Each chamber was filled with 400 ml dechlorinated water. For each metal, nine test containers were allocated, of which eight were for the metal concentrations and one control. Environmental parameters were monitored throughout the 96-hr period. At the termination of each test, mortality data was compiled and the 96-hr $LC_{50}$ values were calculated using the trimmed Spearman-Karber toxicity program. It was observed that the most sensitive species of *Chironomus* larvae was *Gymnometriocnemus mahensis*. The least sensitive form of *Chironomus* larvae examined was *Paramerina minima*. The 96-hr $LC_{50}$ values obtained for *Tanypus complanatus* for Cu, Cd, Hg and Zn were 1.9, 1.35, 0.33 and 28.96 ppm, respectively. Similar results were obtained for other species studies.

Michailova *et al.* (2013) recorded a new species named *Chironomus amissum* from southeastern, Brazil. They described its larva, pupa imago (male) and karyotype. Karyotype belongs to psedothummi cytocomplex with 2n=8 and chromosome arm combination: AE BF CD G. The larval, pupal and adult specimens for morphological analyses were mounted in Euparal medium after being cleared in 10% potassium hydroxide solution. For the karyological analysis, larvae of $4^{th}$ instar were fixed in alcohol/acetic acid-3:1. Preparations of the polytene chromosome were obtained from squashes of salivary gland cells stained with aceto-orcein (Michailova 1989). All preparations are kept in Institute of Biodiversity and Ecosystem Research, Bulgarian Academy of Sciences, Sofia.

Chavan *et al.* (2013) studied the taxonomy of genus *Chironomus* from Balaghat ranges in Beed district of Maharashtra. They illustrate the taxonomic features of *Chironomus circumdatus, Chironomus javanus, Chironomus riparius* and *Chironomus stigmaterus*. Larvae were randomly collected from study area and preserved in 70-80% alcohol. Larvae were kept in 10% KOH for 8-12 hours. After clearing the larvae were passed in distilled water (5 min) then glacial acetic acid (10 min), 2-propanol (15 min), then mounted directly into Canada balsam. Photograph of specimen taken through Olympus camera in Magnus MLXTr microscope at 10x, 40x, 100x magnification. All measurements analyzed through Magnus Pro software. Specimens are deposited in Entomological Research Laboratory, Dr Babasaheb Ambedkar Marathwada University, Aurangabad, Maharashtra, India.

Silveira *et al.* (2013) analyzed the colonization of Chironomidae larvae during the decomposition of *Eichornia azurea* leaves in a lake in southeastern Brazil in two seasons of the year. They found that the feeding activity combined with high larval density is an important factor contributing to the rapid decomposition of the *Eichornia azurea* leaves. They concluded that the succession process along the detritus chain of *Eichornia azurea* was more important in structuring the assemblage of Chironomidae larvae than seasonal variations.

Ebrahimnezhad and Allahemoglobinakhshi (2013) studied the Chironomid larvae of Golpayegan River (Isfahan, Iran). Samples were collected from five sites in the Golpayegan River and identified at generic level using available identification keys. Thirty five genera were identified in four subfamilies including Chironominae (15 genera), Orthocladiinae (13 genera), Tanypodinae (5 genera) and Diamesinae (2 genera). 17 genera of these were reported from the Golpayegan River for the first time.

Hyoung-ho *et al.* (2013) observed the effects of water temperature on development and heavy metal toxicity in two species, *Chironomus riparius* and *Chironomus yoshimatsui* in current era of rapid climate change. These two species exhibit different developmental characteristics and responses to cadmium and lead with temperature. There was a decline in developmental time for both species with temperature. In the acute toxicity test, the 48-hr $LC_{50}$ values for cadmium and lead decreased with temperature for both species. In the chronic toxicity test, emergence rates tended to decrease with temperature, except for when *Chironomus yoshimatsui* was exposed to cadmium.

Midya *et al.* (2013) developed a model on the structure and organization of the polytene chromosome based on the study of *Chironomus striatipennis* Kieffer. A study on the polytene chromosomes obtained from the salivary gland cells of *Chironomus striatipennis* Kieffer indicated that a polytene chromosomes is not a typical ribbon like body formed of a number of thread like chromatin elements aligned side by side having condensed chromatin thread at locations forming bands, simple linear threads of chromatin as interbands and highly extended chromatin threads forming Balbiani rings and puffs at places as conventionally thought of instead a chromosome appears to be a cylindrical body with thin and thick circular bands as elevated regions over a chromosome. The polytene chromosomes prepared from the Chironomid larvae showed the characteristic configurations to support the organization of polytene chromosomes as conjectured. The variable dimensions of bands appeared due to variable amount of condensed chromatin material and intervening interbands also varied in their magnitude of extension. Besides some region of excessive constriction also appeared as waists along a chromosome arm and such differentiation could be possible due to presence of some quasi-fluid substance making the core of the cylindrical chromosome. The multiple chromatin threads in the polytene chromosome appeared to be arranged on the surface line of the long chromosome with coiled or kinky orientation of the thread at identical locations in all the chromatin threads. The chromatin threads were so arranged that the surface of the chromosome appeared smooth laterally except some swelled protrusions at places. The swelled protrusions were the bands and the depressed regions between them were the interbands. The linear core of the cylindrical chromosome appeared to be filled up with a material of quasi-fluid consistency.

Saha and Mazumdar (2013) studied deformities of *Chironomus* sp. larvae as indicator of pollution stress in rice fields of Hooghly District, West Bengal. Morphological deformities of *Chironomus* larvae, particularly in the teeth of the mentum, had been proposed as a bioindicator of sediment quality and environmental stress. Larvae of *Chironomus* sp. were collected from rice fields located at Rishra, Serampore and Khanakul (District- Hooghly, West Bengal, India). At each site physico-chemical parameters of water and sediment were recorded. Field data exhibit high incidence of deformity in Rishra compared with Serampore and Khanakul. The agronomic practices revealed an excessive application of pesticide in the rice field for better yield. The rice field ecosystems thus contaminated by pesticides and cause toxic effects in *Chironomus* sp. despite those other main contaminating agent i.e. industrial effluents in the adjoining rice fields.

Ree Han (2013) described six new and two newly recorded species of Chironomidae in Korea. Adult Chironomids were collected by light trap operations, sweeping grasses with insect net and aspiration of light attracted adults at various localities. All collected specimens were slide mounted and identified. Six species new to science were found new and named: *Rheotanytarsus pseudopentapoda* n. sp., *Cricotopus byeonsanensi* n. sp., *Cricotopus parajogantertius* n. sp., *Limnophyes bukhadecimus* n. sp., *Limnophyes simensis* n. sp. and *Orthocladius jeongnungensis* n. sp.. Two species, *Ainuyusuroka tuberculatum* (Tokunga) and

*Pseudosmittia nishiharaensis* (Sasa and Hasegawa) are found for the first time in Korea. All species are fully described with illustrations. This is the first report of the genus *Ainugurusika* in Korea.

Xing Li and Xin-hua Wang (2014) recorded a new species, *Metriocnemus calcaneum* sp. n. from China. They also provided the key to the males of this species. The material was mounted on slides in Canada balsam, following the procedure outlined by Saether (1969). The types and other examined material are housed in the College of Life Sciences, Nankai University, China.

Siri and Brodin (2014) described a new species, *Rheochlus latisetus* sp. nov. from male and female adults from Argentina. They also illustrated and emendated three previously known species of *Rheochlus* Brundin, 1966. They also presented a cladistic analysis of the genus *Rheochlus* within the tribe Podonomini. Specimens of cleared *Rheochlus latisetus* sp. nov., the holotype of *Rheochlus prolongatus*, paratypes of *Rheochlus insignis* and *Rheochlus wirthi* were slide mounted in Canada balsam. Holotype, allotype and paratypes of *Rheochlus latisetus* were deposited in the collection of the Museo de La Plata, Argentina (MLPA). Material of *Rheochlus insignis, Rheochlus prolongatus, Rheochlus wirthi* and one paratype of *Rheochlus latisetus* are deposited in the Swedish Museum of Natural History, Stockholm, Sweden (SMNH). For cladistic analysis 32 different characters were coded as non additive and analyzed with program TNT version 1.1 (Goloboff *et al.* 2008a) under optimal criteria.

Baranov *et al.* (2014) erected a new genus of Podonominae, *Palaeoboreochlus* Baranov *et* Anderson n. gen. based on *Palaeoboreochlus inornatus* Baranov *et* Anderson n. sp. described from a male found in Late Eocene Rovno amber from Ukraine. The new genus grouped with *Boreochlus* Edwards in the tribe Boreochlini. The fossil was studied using a Nikon Optiphot 2 microscope. The drawings were made with the aid of drawing tube and processed in Adobe Photoshop CS5. The photos were taken with the aid of a Leica DM 400B LED microscope using a Leica DFC 450C compact camera. The type is deposited in the I. I. Schmalhausen Institute of Zoology, National Academy of Science of Ukraine, Kiev (SIZK), Ukraine.

Gunderina LI (2014) developed a design of molecular markers for identification of species of the genus *Chironomus*. Accurate identification and differentiation of species of genus *Chironomus* based on their morphological features is a difficult problem. Unambiguous species identification by means of molecular markers is possible at any stage of the life cycle. Polymerase chain reaction (PCR) with species specific primers was used to develop molecular markers (amplicons) for identification of *Chironomus piger, Chironomus dorsalis* and *Chironomus pseudothummi.* Nucleotide sequences of the internal transcribed spacer region (ITS) of the locus coding for ribosomal RNA were used to design species specific primers for these target species. Each of the species specific primer pairs yielded species specific amplicons only with the DNA of target species: *Chironomus piger, Chironomus dorsalis* and *Chironomus pseudothummi.* Test PCRs with the DNA of eighteen *Chironomus* species confirmed the specificity of the primers obtained. The molecular markers produced in PCR with the designed species specific primers permit reliable identification of *Chironomus piger, Chironomus dorsalis* and *Chironomus pseudothummi* and their differentiation from other species of genus *Chironomus.*

Paul N *et al.* (2014) described Immature and adult stages of *Monopelopia mongpuense* sp. n. from phytotelmata of Cedrus deodara in Darjeeling were described along with biological notes. Key to the adult males of all species of the genus *Monopelopia* Fittkau was also presented. This genus was recorded for the first from Indian subcontinent.

Petrova NA and Zhirov SV (2014) prepared a report on characteristics of the karyotypes of three subfamilies (Tanypodinae, Diamesinae and Prodiamesinae) of Chironomids of the world fauna. Karyotypes and characteristics features of the polytene chromosomes are described for 55 species of the three subfamilies; this is almost three times as many species as were included in the previous reviews published in 1980s.

Sulistiyarto B *et al.* (2014) developed a production technique of Chironomidae larvae in floodplain waters for fish food. Chironomid larvae are a natural food that has a nutrient suitable for freshwater fish needs. Bloodworm utilization as fish feed in aquaculture is still limited, due to limited production. This study showed that the characteristics of best location for the production of bloodworm were in the waters of the swamp forest covered tree canopy with water depth of 1 to 2 m. Type of a good substrate for the production of bloodworm was coconut fiber. The substrate was placed in a horizontal position at the bottom. Bloodworm production in this experiment was 0.938 g dry weight/m$^2$ or 32.83 g wet weight/m$^2$.

Ilkova *et al.* (2014) reported genome instability of *Chironomus riparius* Mg. from polluted water basins in Bulgaria. Larvae of *Chironomus riparius* Mg. collected from two polluted water basins in Bulgaria plus controls reared in the laboratory were studied. High concentrations of the heavy metals Pb, Cu and Cd were recorded in the sediments of the polluted stations. Marked somatic structural chromosome aberrations were found in *Chironomus riparius* salivary polytene chromosomes from the field stations and their frequency was significantly higher (p<0.01) compared to control. The observed somatic chromosome changes are discussed as a response of the Chironomid genome to aquatic pollution. A new cytogenetic index based on the number of aberrations found in larvae from polluted regions in comparison with the control was applied to the data to more easily evaluate the degree of heavy metal pollution in aquatic ecosystems. Their study of a polluted site near the river Chaya showed that the somatic index was very high at 3.35 for 2010 and 11.66 for 2013 compared to 0.5 in the control. The cytogenetic index was effective in showing that all studied sites were highly polluted in comparison with the control. To determine the mechanism involved in the concentration of aberration breakpoints within specific regions of Chironomid polytene chromosome the FISH method was applied. The localization of a transposable element TFB1 along the polytene chromosomes of *Chironomus riparius* was analyzed and the sites of localization were compared with breakpoints of chromosome aberrations. A significant correlation ($p<0.05$) was found which shows that most of the aberrations do not appear randomly but are concentrated in sites rich in transposable elements.

Boggero *et al.* (2014) found a new species of *Pseudomittia* Edwards 1932, *Pseudosmittia fabioi*, from Sardinia (Palearctic region). The generic diagnosis is emended based on characters found in the new species. This species is characterized by a triangular and point with microtrichia at apex, an inferior vulsella characterized by a rounded lobe with a setose accessory lobe adrepressed to the genooxite, a well developed tooth like projection on the outer margin of gonostylus, the lack of acrostichals and an antenna with 6 flagellomeres in female.

Fard MS *et al.* (2014) prepared a report on Chironomidae larvae as aquatic amphibian food. Chironomid larvae considered as an important food source for amphibians. In their study three species of Chironomidae (*Baeotendipes noctivagus, Benthalia dissidens* and *Chironomus riparius*) were identified in 23 samples of larvae from Belgium, Poland, Russia and Ukraine provided by a distributor in Belgium. They evaluated the suitability of these samples as amphibian food based on four different aspects: the likelihood of amphibian pathogens spreading risk of heavy metals accumulation in amphibians, nutritive value and risk of spreading zoonotic bacteria (*salmonella, Campylobacter* and *Enterobacteriaceae*). They found neither zoonotic bacteria nor the amphibian pathogens in these samples. Proximal nutritional analysis revealed that Chironomidae larvae are consistently high in protein but more variable in lipid content. Accordingly variations in the lipid:protein ration can affect the amount and pathway of energy supply to the amphibians. They showed that Chironomidae larvae may not be recommended as single diet item for amphibians.

Nandi *et al.* (2014) give evidence that Chironomid midges may act as allergens from two species from West Bengal, Kolkata, India. This study was done to access two common Chironomid species *Chironomus circumdatus* and *Polypedilum nubifer* for their sensitizing potential as an allergen in atopic patients and controls. Following preparation of allergenic extracts of the two Chironomid species separately, 198 atopic patients attending an allergy clinic and 50 age matched controls were tested along with a routine panel of allergen to assess sensitization. The skin prick test (SPT) results revealed that 189 of the 198 patients (95.4%) demonstrated sensitization to both the Chironomid species. Higher levels of total IgE was observed in atopic subjects than in the control group. The result suggested that the Chironomid midges *Chironomus circumdatus* and *Polypedilum nubifer* can elicit sensitization in humans. A potential risk for allergic reaction by susceptible individuals exists due to these Chironomid species, owing to their abundance and chance of contact with human beings. Furthermore studies may be initiated to characterize the nature of the allergens and to assess their clinical relevance.

Saha D and Mazumdar A (2014) found that the incidence of deformities in the head capsule of Chironomid larvae are considered as indicators of environmental pollution such as heavy metals, pesticides, but as well by organic contaminants. Deformities of *Chironomus sp.* (Diptera: Chironomidae) larvae have been proposed as a bioindicator of sediment quality and environmental stress. Larvae of *Chironomus* sp. were collected from rice fields located at Rishra, Serampore and Khanakul (District- Hooghly, West Bengal, India). Significant differences were observed in deformities between three sites. The agronomic practices revealed an overuse of pesticides in the rice field for better productivity. Samples obtained from Rishra exhibited high incidence of deformity compared with Serampore and Khanakul. Rishra was polluted area and various types of industrial effluents discharge into the rice field periodically. Beside industrial effluents the farmer uses large amount of pesticide in the rice field. Laboratory experiments demonstrated a correlation between field concentration of pesticide and occurrence of deformed larvae. The occurrence of higher frequencies of deformities in Rishra indicated that the frequency of deformities was apparent with increase of environmental pollutants.

# REFERENCES


**Al-Saffar AT.** Larval mouthpart deformities in *Chironomus annularis* Meigen from Al-Hammar marsh, Southern Iraq and Tanjero River, Kurdistan, Northern Iraq. *Nature Iraq*, 2007.

**Al-Shami S, Rawi CSM, Nor SAM, Ahmad AH and Ali A.** Morphological deformities in *Chironomus* spp. larvae as a tool for impact assessment of anthropogenic and environmental stresses on three rivers in the Juru River system, Penang, Malaysia. *Environ. Entomol.* 39 (1): 210-222, 2010.

**Armitage PD, Cranston PS and Pinder LCV.** The Chironomidae: Biology and Ecology of non-biting midges. *Chapman and Hall*, ISBN: 041245260X, 1995.

**Ashe P, Murray A and Reiss F.** The zoogeographical distribution of Chironomidae (Insecta: Diptera). *Annls. Limnol.* 23 (1): 27-60, 1987.

**Baranov V, Anderson T & Perkovsky E.** A new genus of Podonominae (Diptera: Chironomidae) in Late Eocene Rovno amber from Ukraine. *Zootaxa* 3794 (4): 581–586, 2014.

**Belikov S and Wieslander L.** Improved preservation of chromatin structure in ethanol fixed cells. *Nucleic acids res.* 22 (10): 1928-1929, 1984.

**Bergtrom G, Laufer H and Rogers R.** Fat body: A site of hemoglobin synthesis in *Chironomus thummi* (Diptera). *J. Cell Biol.* 69: 264-274, 1976.

**Bhaduri S, Sarkar P, Ghosh C and Midya T.** Response of the Chironomid larvae to the environmental condition: A study on the polytene chromosomes of *Chironomus striatipennis* (Kieffer). *The Ecoscan* 5: 75-80, 2011.

**Boggero A, Zaupa S & Rossaro B.** *Pseudosmittia fabioi* sp. n., a new species from Sardinia (Diptera: Chironomidae, Orthocladiinae). *J. Entomol. Acaro.l Res.* 46: 1892, 2014.

**Brackenbury J.** Locomotory modes in the larva and pupa of *Chironomus plumosus* (Diptera, Chironomidae). *J. Insect Physiol.* 46: 1517-1527, 2000.

**Casas JJ.** The effect of diet quality on growth and development of recently hatched larvae of *Chironomus gr. plumosus. Limnetica* 12 (1): 1-8, 1996.

**Chaudhuri PK, Hazra N and Alfred JRB.** A checklist of Chironomid midges (Diptera: Chironomidae) of the Indian subcontinent. *Oriental Insects* 35: 335-372, 2001.

**Chavan RJ, Gaikwad AM, Shinde LV and Shonune BV.** Studies on taxonomy of genus *Chironomus* (Meigen 1803) (Insecta: Chironomidae) from Balaghat Ranges in Beed district of Maharashtra. *The Ecoscan* 4: 117-121, 2013.

**Correia LCS, Susana TS and Michailova P.** *Chironomus amissum* sp. n. (Diptera. Chironomidae) from southeastern Brazil. *Biota. Neotrop.* 13 (4): 133-138, 2013.

**Cranston PS.** A new genus and species of Chironominae with wood-mining larvae. *Aust. J. Entomol.* 45: 227-234, 2006.

**Donna JA, Michel RN, David AZ, Carol G, Eugen W and Thomas MJ.** Left handed Z-DNA in bands of acid fixed polytene chromosomes. *Proc. Natl. Acad. Sci. USA* 80: 4344-4348, 1983.

**Dutta TK, Ali A, Majumdar A and Chaudhuri PK.** Chironomid midges of *Harnischia* complex (Diptera: Chironomidae) from the duars of the Himalayas, India. *Eur. J. Entomol.* 93: 263-279, 1996.

**Ebau W, Rawi CS, Din Z and Al-Shami A.** Toxicity of cadmium and lead on tropical midge larvae, *Chironomus kiiensis tokunga* and *Chironomus javanus kieffer. Asian Pac. J. of Trop. Biomed.* 2 (8): 631-634, 2012.

**Ebrahimnezhad M and Allahemoglobinakhshi E.** A study on Chironomidae larvae of Golpayegan River (Isfahan-Iraq) at generic level. *Iran. J. of Sci. and Technol.* A1: 45-52, 2013.

**Edstrom JE and Beermann W.** The base composition of nucleic acids in chromosomes, puffs, nucleoli and cytoplasm of *Chironomus* salivary gland cells. *J. Cell Bio.*14, 1962.

**Eggermont H and Heiri O.** The Chironomid-temperature relationship: expression in nature and palaeoenvironmental implications. *Biol. Rev.* 87: 430-456, 2012.

**Ewer RF.** On the function of hemoglobin in *Chironomus. J.E.E.* 18 (3), 1942.

**Fard MS, Pasmans F, Adriaensen C, Laing GD, Janssens GPJ and Martel A.** Chironomidae bloodworms larvae as aquatic amphibian food. *Zoo Biol.* 33: 221-227, 2014.

**Gunderina LI.** Design of molecular markers for identification of species of the genus *Chironomus* (Diptera, Chironomidae). *Entomol. Rev.* 94(1): 140-148, 2014.

**Guryev V, Makarevitch I, Blinov A & Martin J.** Phylogeny of the genus *Chironomus* (Diptera) inferred from DNA sequences of mitochondrial Cytochrome b and Cytochrome oxidase I. *Mol. Phylo and Evo.* 19(1): 9-21, 2001.

**Habashy MM.** Culture of Chironomid larvae under different feeding systems. *Egypt. J. aquat. Res.* 31 (2): 403-418, 2005.

**Halpern M, Broza YB, Mittler S, Arakawa E and Broza M.** Chironomid egg masses as a natural reservoir of *Vibrio cholera* Non-01 and Non-0139 in freshwater habitats. *Microbial ecol.* 47: 341-349, 2004.

**Halpern M, Gancz H, Broza M and Kashi Y.** *Vibrio cholerae* hemagglutinin/ protease degrades Chironomid egg masses. *Appl. Environ. Microbiol.* 69 (7): 4200-4204, 2003.

**Halpern M, Gasith A, Bresler VM and Broza M.** The protective nature of *Chironomus luridus* larval tubes against copper sulphate. *J. Insect Sci.* 2 (8), 2002.

**Henriques-Oliveira AL, Neissimian JL and Dorville LFM.** Feeding habits of Chironomid larvae (Insecta: Diptera) from a stream in the Floresta da Tijuca, Rio De Janeiro, Brazil. *Braz. J. Biol.* 63 (2): 269-281, 2003.

**Henry R and Santos CM.** The importance of excretion by *Chironomus* larvae on the internal loads of nitrogen and phosphorus in a small eutrophic urban reservoir. *Braz. J, Biol.* 68 (2): 349-357, 2008.

**Ilkova J, Michailova P, Thomas A & White K.** genome instability of *Chironomus riparius* Mg. (Diptera, Chironomidae) from polluted water basins in Bulgaria. *Ecologia Balkanica.* 5: 1-8, 2014.

**Kiknadze II, Istomina AG, Wulker WF and Vallenduuk HJ.** The karyotype of *Chironomus uliginosus* Keyl. *Becmhuk BOTUC* 14 (1): 22-28, 2010.

**Lencioni V, Marziali L and Rossaro B.** Diversity and distribution of Chironomids (Diptera, Chironomidae) in pristine Alpine and pre-Alpine springs (Northern Italy). *J. Limnol.* 70 (1): 106-121, 2011.

**Lopez CC, Nielsen L and Edstrom JE.** Terminal long tandem repeats in chromosomes from *Chironomus pallidivittacus. Mol. Cell Biol.* 16 (7): 3285-3290, 1996.

**M Ebrahimnezhad and F Fakhri.** Taxonomic study of Chironomidae larvae of Zayandehrood River, Iran and effects of selected ecological factors on their abundance and distribution. *IJST* 29, 2005.

**Madalena CRG, Diez JL and Gorab E.** Chromatin structure of ribosomal RNA genes in Dipterans and its relationship to the location of nucleolar organizers. *Poison* 7 (8): e44006, 2012.

**Majumdar TN and Gupta N.** Acute and chronic toxicity of copper on aquatic insect *Chironomus ramosus* from Assam, India. *J. Environ. Biol.* 33: 139-142, 2012.

**Marino M, Kasper de B, Michel A, Maxine B, Martijs JJ, Michiel HSK and Wim A.** Response of the non-biting midge *Chironomus riparius* to multigeneration toxicant exposure. *Environ. Sci. Technol.* 46: 12105-12111, 2012.

**Martinez JL, Sanchez-Elsner T, Morcillo G and Diez JL.** Heat shock regulatory elements are present in telomeric repeats of *Chironomus thummi. Nucleic acids res.* 29 (22): 4760-4766, 2001.

**Michailova P, Iikova J, Petrova N and White K.** Rearrangements in the salivary gland chromosomes of *Chironomus riparius* Mg. (Diptera: Chironomidae) following exposure to Lead. *Caryologia* 54: 349-363, 2001.

**Michailova P, Petrova N, Sella G, Bovero S, Ramella L, Regoli F and Zelano V.** Genotoxic effects of chromium on polytene chromosomes of *Chironomus riparius Meigen* (Diptera: Chironomidae). *Caryologia* 54 (1): 59-71, 2001.

**Michailova P, Sella G and Petrova N.** Chironomids (Diptera) and their salivary gland chromosomes as indicators of trace-metal genotoxicity. *Ital. J. Zool.* 79 (2): 218-230, 2012.

**Michailova P, Szarek-Gwiazda E, Kownacki A and Warchalowska-Sliwa E.** Biodiversity of Chironomidae (Diptera) and genome response to trace metals in the environment. *Pesticides* 1-4: 41-48, 2011.

**Michailova P, Warchalowska-Sliva E, Krastanov B and Kowncki A.** Cytogenetic variability in species of genus *Chironomus* from Poland. *Caryologia* 58 (4): 345-358, 2005.

**Midya T, Bhaduri S and Sarkar P.** Failure in somatic pairing of 4th chromosome in *Chironomus striatipennis* Kieffer. *The Bioscan* 7 (2): 321-324, 2012.

**Midya T, Sarkar P and Bhaduri S.** Effect of Lead as heavy metal in aquatic habitat to promote polymorphism of polytene chromosomes in *Chironomus striatipennis* Kieffer (Diptera: Chironomidae). *The Ecoscan* 1: 173-178, 2012.

**Midya T, Sarkar P, Bhaduri S and Ghosal SK.** A model on the structure and organization of the polytene chromosome based on the study on *Chironomus striatipennis* Kieffer (Diptera: Chironomidae). *The bioscan* 8(1): 21-21, 2013.

**Mo H, Yoo D, Bae Y and Cho K.** Effects of water temperature on development and heavy metal toxicity change in two midge species of *Chironomus riparius* and *C. yoshimatsui* in an era of rapid climate change. *Entomol. Res.* 43: 123-129, 2013.

**Morais SS, Molozzi J, Viana TH and Callisto M.** Diversity of larvae of littoral Chironomidae and their role as bioindicators in urban reservoirs of different trophic levels. *Braz. J. Biol.* 70 (4): 995-1004, 2010.

**Morcillo G and Diez JL.** Telomeric puffing induced by heat shock in *Chironomus thummi. J. Biosci.* 2: 247-257, 1996.

**Nandi S, Aditya G, Chowdhury I & Saha GK.** Chironomid midges as allergens: evidence from two species from West Bengal, Kolkata. *India. Indian J Med Res.* 129: 921-926, 2014.

**Nazarova LB, Riss HW, Kahlheber A and Werding B.** Some observation of buccal deformities in Chironomid larvae from the Cienga Grande De Santa Marta, Colombia. *Caldasia* 26 (1): 275-290, 2004.

**Ozkan N and Camur-Elipek B.** The dynamics of Chironomidae larvae and the water quality in Meric River (Edirne/Turkey). *Tiscia* 55: 49-54, 2006.

**Ozkan N, Camur-Elipek B and Moubayed BJ.** Ecological analysis of Chironomid larvae (Diptera, Chironomidae) in Ergene River basin (Turkish Thrace). *Turk. J. Fish. Aquat. Sci.* 10: 93-99, 2010.

**Paul N, Hazra N & Mazumdar A**. *Monopelopia mongpuense* sp. n., a phytotelmata midge from sub–Himalayan region of India (Diptera: Chironomidae: Tanypodinae). *Zootaxa* 3802 (1): 122-130, 2014.

**Pery ARR, Mons R, Flammarion P, Lagadic L and Garric J**. A modeling approach to link food availability, growth, emergence, and reproduction for the midge *Chironomus riparius*. *Environ. Toxicol. Chem.* 21 (11): 2507-2513, 2002.

**Petrova NA and Zhirov SV**. Polytene chromosomes of salivary glands of Chironomidae (Diptera: Chironomidae) from the Wrangel Island (Russia). *Comparative Cytol.* 2 (2): 127-13, 2008.

**Petrova NA and Zhirov SV**. Characteristics of the karyotypes of three subfamilies of Chironomids (Diptera, Chironomidae: Tanypodinae, Diamesinae, Prodiamesinae) of the world fauna. *Entomol. Rev.* 94(2): 157-165, 2014.

**Petrova NA, Zhirov SV, Maria VH and Karine VH**. Cytotaxonomy and Morphology of Chironomid larvae (Diptera, Chironomidae) in Armenia. *Int. J. Med. and Biol. Sci.* 6, 2012.

**Pfenninger M and Nowak C**. Reproductive isolation and ecological niche partition among larvae of the morphologically cryptic sister species *Chironomus riparius* and *Chironomus piger*. *Poison* 3 (5): e2157, 2008.

**Postma JF, Buckert-de Jong MC, Staats N and Davids C**. Chronic toxicity of Cadmium to *Chironomus riparius* at different food levels. *Arch. Environ. Contam. Toxicol.* 26: 143-148, 1994.

**Ree H**. Six new and two newly recorded species of Chironomidae (Insecta: Diptera) in Korea. *Entomol. Res.* 43: 322-329, 2013.

**Rico E and Quesada A**. Distribution and ecology of Chironomids (Diptera, Chironomidae) on Byers Peninsula, Maritime Antarctica. *Antarct. Sci.* 25 (2): 288-291, 2013.

**Rovira C, Beerman W and Edstrom JE**. A repetitive DNA sequence associated with the centromeres of *Chironomus pallidivittatus*. *Nucleic acids res.* 21 (8): 1755-1781, 1991.

**Saether OA**. Phylogeny of the subfamilies of Chironomidae (Diptera). *Syst. Entomol.* 25: 393-403, 2000.

**Saha D & Mazumdar A**. Deformities of *Chironomus* sp. larvae (Diptera: Chironomidae) as indicator of pollution stress in rice fields of Hooghly District, West Bengal. *JTBSRR* 2 (2): 44-54, 2013.

**Saha D & Mazumdar A.** The use of head capsule deformities in Chironomid larvae (Diptera: Chironomidae) to assess environmental pollution in rice fields of Hooghly District, West Bengal, India. *IJISET* 1(5): 1-8, 2014.

**Sanseverino AM and Nessimian JL.** The food of larval Chironomidae (Insecta, Diptera) in submerged litter in a forest stream of the Atlantic Forest (Rio de Janeiro, Brazil). *Acta Limnol. Bras.* 20 (1): 15-20, 2008.

**Silva FL, Ruiz SS, Bochini GL and Moreira DC.** Functional feeding habits of Chironomidae larvae in a lotic system from Midwestern region of Sao Paulo State, Brazil. *Panam. J. Aquat. Sci.* 3 (2): 135-141, 2008.

**Siri A and Brodin Y.** Cladistic analysis of *Rheochlus* and related genera, with description of a new species (Diptera: Chironomidae: Podonominae). *Acta Entomol. Musei Nat. Pra.* 54 (1): 361-375, 2014.

**Soon-Mi Lee, Se-Bum Lee, Chul-Hwi Park and Jinhee Choi.** Expression of heat shock protein and hemoglobin genes in *Chironomus tentans* larvae exposed to various environmental pollutants: A potential biomarker of freshwater monitoring. *Chemosphere* 65: 1074-1081, 2006.

**Sulistiyarto B, Christiana I and Yuiintine.** Developing production technique of bloodworm (Chironomid larvae) in floodplain waters for fish food. *Int. J. Fish. Aquac.* 6(4): 39-45, 2014.

**Thompson PE and Bowen JS.** Interactions of differentiated primary sex factors in *Chironomus tentans*. *Genetics* 70: 491-493, 1971.

**Vedamanikam VJ, Bandara A and Boncoeur P.** A study on the effects of four heavy metals on five species of *Chironomus* larvae present on Mahe Island of Seychelles. *Toxicol. Environ. Chem.* DOI: 10.1080/02772248.2013.862388.

**Wulker WF, Kiknadze II and Istomia AG.** Karyotypes of *Chironomus* Meigen (Diptera: Chironomidae) species from Africa. *Comp. Cytogenetics* 5 (1): 23-48, 2011.

**Xin Qi, Xiaolong Lin and Xinhua Wang.** A new species of the genus *microtendipes* Kieffer, 1915 (Diptera, Chironomidae) from Oriental China. *Zookeys* 212: 81-89, 2012.

**Xing Li and Xin-hua Wang.** New species and records of *Metriocnemus* van der Wulp s. str. from China (Diptera, Chironomidae). *Zookeys* 387: 73-87, 2014.

# ABBREVIATIONS

| | |
|---|---|
| ANOVA | Analysis of variance |
| BOD | Biological oxygen demand |
| bp | Base pair |
| BR | Balbiani ring |
| cc | Cubic centimeter |
| COD | Chemical oxygen demand |
| DDT | Dichlorodiphenyltrichloroethane |
| DNA | Deoxyribonucleic acid |
| *et al.* | and others |
| Hz | Hertz |
| Kb | kilo byte |
| $LC_{50}$ | Lethal concentration, 50% |
| NOR | Nucleolar organizer region |
| ppm | Part per million |
| RNA | Ribonucleic acid |
| sp. | Species |
| viz. | namely |

www.ingramcontent.com/pod-product-compliance
Lightning Source LLC
LaVergne TN
LVHW090857240726
843527LV00049B/69